Time and Becoming in Nietzsche's Thought

Continuum Studies in Continental Philosophy
Series Editor: James Fieser, University of Tennessee at Martin, USA

Continuum Studies in Continental Philosophy is a major monograph series from Continuum. The series features first-class scholarly research monographs across the field of Continental philosophy. Each work makes a major contribution to the field of philosophical research.

Adorno's Concept of Life, Alastair Morgan
Badiou, Marion and St Paul, Adam Miller
Being and Number in Heidegger's Thought, Michael Roubach
Deleuze and Guattari, Fadi Abou-Rihan
Deleuze and the Genesis of Representation, Joe Hughes
Deleuze and the Unconscious, Christian Kerslake
Deleuze, Guattari and the Production of the New, edited by Simon O'Sullivan and Stephen Zepke
Derrida, Simon Morgan Wortham
Derrida and Disinterest, Sean Gaston
The Domestication of Derrida, Lorenzo Fabbri
Encountering Derrida, edited by Simon Morgan Wortham and Allison Weiner
Foucault's Heidegger, Timothy Rayner
Gadamer and the Question of the Divine, Walter Lammi
Heidegger and a Metaphysics of Feeling, Sharin N. Elkholy
Heidegger and Aristotle, Michael Bowler
Heidegger and Philosophical Atheology, Peter S. Dillard
Heidegger Beyond Deconstruction, Michael Lewis
Heidegger, Politics and Climate Change, Ruth Irwin
Heidegger's Contributions to Philosophy, Jason Powell
Heidegger's Early Philosophy, James Luchte
The Irony of Heidegger, Andrew Haas
Levinas and Camus, Tal Sessler
Merleau-Ponty's Phenomenology, Kirk M. Besmer
Nietzsche, Nihilism and the Philosophy of the Future, edited by Jeffrey Metzger
Nietzsche's Ethical Theory, Craig Dove
Nietzsche's Thus Spoke Zarathustra, edited by James Luchte
The Philosophy of Exaggeration, Alexander Garcia Düttmann
Sartre's Phenomenology, David Reisman
Who's Afraid of Deleuze and Guattari? Gregg Lambert
Žižek and Heidegger, Thomas Brockelman

Time and Becoming in Nietzsche's Thought

Robin Small

continuum

Continuum International Publishing Group
The Tower Building
11 York Road
London SE1 7NX

80 Maiden Lane
Suite 704
New York, NY 10038

www.continuumbooks.com

British Library Cataloguing-in-Publication Data
A catalogue record for this book is available from the British Library.

ISBN: PB: 978-1-4411-4794-3

Library of Congress Cataloging-in-Publication Data
Small, Robin, 1944-
Time and becoming in Nietzsche's thought / Robin Small.
p. cm.
Includes bibliographical references and index.
ISBN 978-1-4411-4794-3
1. Nietzsche, Friedrich Wilhelm, 1844-1900. 2. Time. 3. Becoming (Philosophy) I. Title.
B3318.T5S63 2010
115.092–dc22

2009028818

Typeset by Newgen Imaging Systems Pvt Ltd, Chennai, India
Printed and bound in Great Britain by the MPG Books Group

This book is dedicated to Daniel Clough: 'The child is innocence and forgetting, a new beginning, a game, a self-rolling wheel, a first movement, a sacred "Yes".' (Nietzsche, Thus Spoke Zarathustra, *'On the Three Metamorphoses')*

Contents

List of Figures

Preface

Interpretations of Friedrich Nietzsche's philosophy have never been hard to find. Only lately, however, has it been widely recognized that careful and critical thought is just as necessary in approaching his ideas as in dealing with the central figures of the Western philosophical tradition. In this book I offer an extended discussion of one theme found in his work, and especially in *Thus Spoke Zarathustra*: the related concepts of time and becoming. Not everyone shares my belief that Nietzsche makes a profound and important contribution to this topic. My hope is that doubters may reconsider on reading this book. For those who already agree, my treatment will aim first at opening up unfamiliar aspects of Nietzsche's thinking and then, more boldly, at proposing a new interpretation of his conception of temporality.

We associate Nietzsche with the doctrine of eternal return, especially in view of the high claims he made for its importance, and yet there is far more to his thought about time and becoming than this concept alone. For that reason, this book is not primarily about the eternal return. Although it figures in several chapters, many of its aspects will not be touched on. However, an exploration of Nietzsche's thinking on becoming and time should open the way to a more adequate treatment of eternal recurrence. In saying this, I am signalling a particular line of interpretation: I take the doctrine to be *about time*. This is the approach of some other writers, but more common is an alternative that is oriented towards issues of human personality and forms of life, in which the thought figures as a symbol, myth, ideal or heuristic principle. That position is stated succinctly by Alexander Nehamas: 'The eternal recurrence is not a theory of the world but a view of the self.'[1] My belief is that this idea turns its back on some of Nietzsche's most original and challenging ideas. Supporting that criticism, however, means showing in some detail what these ideas are. That is what this book is intended to achieve.

These contrasted approaches are often supposed to correspond to different ways of reading Nietzsche's texts. The claim is that regarding the doctrine as being about the world implies taking it in a literal sense, whereas regarding it as about 'the self' means reading it metaphorically. I see no good reason to accept these associations. It is true that Nietzsche is unusual among philosophers in his reliance on literary modes of presentation. Yet we need to bear in mind his insistence on the metaphorical character of our ordinary concepts, which undermines the whole distinction between the metaphorical and literal, or at least makes it of little use except in deciphering the plainly symbolic elements in his writing. I will argue that Nietzsche's figurative and literary presentations lend themselves, in surprising detail, to the debates about time and becoming that have engaged other thinkers. The hidden complexities of his metaphors mirror the complexities of the corresponding conceptual problems, suggesting that each may usefully throw light on the other.

The most novel aspect of my interpretation will be the thesis that Nietzsche presents not one, but several quite different models of time corresponding to different 'forms of life'. Has this not been noticed by previous commentators? In one particular area, yes. Discussion of the eternal recurrence often raises the idea of a personal transformation which would bring with it a new view of the world – including a new experience of time. But this notion has not been elaborated in the way I set out to do here. That is a central task of this book, and it is my hope that opening up perspectives on temporality will help in understanding many other aspects of Nietzsche's thought.

Acknowledgements

Work towards this book has been made easier and happier by the support and encouragement received from colleagues at the University of Auckland and elsewhere during the writing process, and in particular from Babette Babich, Gabriele Lakomski, Jim Marshall and Lynda Stone. Discussions with fellow scholars, especially Christoph Landerer and Julian Young, helped to develop some of the ideas here. John Mandalios and Paul Loeb each read the complete draft and offered valuable suggestions for improvement in a number of places. Finally, Anna Small helped with preparing the text figures in Chapter 3. My sincere thanks to all these people.

Chapters 3, 4 and 5 include material which appeared as 'Zarathustra's Gateway', in *History of Philosophy Quarterly* 15.1 (1998), 79–98, and 'Zarathustra's Four Ways: Structures of Becoming in Nietzsche', in *British Journal for the History of Philosophy* 9.1 (2001), 83–107. Chapter 7 includes material published in 'Disturbing Thoughts and Eternal Perspectives: Some Uses of Symbolism in Nietzsche', in *New Nietzsche Studies* 3.3/4 (1999): 29–44. Permission to reprint is gratefully acknowledged.

A Note on Translations and Sources

In quoting from the published works of Nietzsche and from the notes included in *The Will to Power*, I have used the translations of R. J. Hollingdale and Walter Kaufmann, and in quoting from Nietzsche's Basel lectures on the pre-Platonic philosophers, the English translation of Greg Whitlock, in both cases with some modifications. Chapter titles and section numbers are given for Nietzsche's works to enable reference to any edition. Other translations are my own, unless otherwise indicated.

The following abbreviations are used in references:

KGW Nietzsche, *Kritische Gesamtausgabe: Werke.* Edited by Giorgio Colli and Mazzino Montinari. 36 vols. Berlin: Walter de Gruyter, 1963–.

KSA Nietzsche, *Kritische Studienausgabe: Sämtliche Werke.* Edited by Giorgio Colli and Mazzino Montinari. 15 vols. Berlin: Walter de Gruyter, 1980.

KSB Nietzsche, *Kritische Studienausgabe: Sämtliche Briefe.* Edited by Giorgio Colli and Mazzino Montinari. 8 vols. Berlin: Walter de Gruyter, 1986.

Introduction

The subject of this book is Friedrich Nietzsche's contribution to an understanding of time, that is, his approach to questions about the nature of becoming, and about past, present and future. Nietzsche is known as one of the most important of Western thinkers, but usually for other reasons. He is valued for his radical critique of metaphysics, morality and religion, or for his sharp diagnoses of modern culture, or for his aphoristic insights into human nature. My aim is to balance these judgements by exposing another side of Nietzsche's achievement. I intend to show that his thought about time and becoming is just as significant. It addresses issues that have concerned philosophy from the beginning to the present day, and does so along highly original lines. For that reason, it deserves close attention and critical reflection, no less than his other ideas.

First and foremost, Nietzsche is a philosopher of becoming. It is not just that he insists on the reality of change, coming to be and passing away, and criticizes any view of the world that leaves these out or explains them away. He upholds a world of continual change and transformation, and the illusoriness of stability and permanence. So too, antagonism to any form of Platonism is a constant theme in his writing, as an early declaration makes clear: 'My philosophy *inverted Platonism*: the further away from real being, the purer, finer, better it is.'[1] This attitude never altered. It is opposed not only to the idea of a 'true world' beyond that of appearance, but also to any belief in a timeless reality. In this sense, Nietzsche is in accordance with the Parmenidean and Platonist identification of true reality with absence of change. Since timelessness is a traditional hallmark of genuine being, a reaffirmation of appearance will also be a reaffirmation of becoming. Conversely, an affirmation of becoming is a vindication of appearance, as opposed to the supposed 'true' world of metaphysics.

It is Heraclitus in whose company, Nietzsche wrote, he felt 'altogether warmer and better than anywhere else'[2] and who above all stands as a precursor for his thinking about becoming and being. Aristotle noted that 'the view is actually held by some that not merely some things but all things in the world are in motion and always in motion, though we cannot apprehend the fact by sense-perception'.[3] This is the teaching of Heraclitus, which I will be calling the doctrine of *absolute becoming*. Aristotle considered it a paradoxical view, and it is certainly one of the boldest philosophical claims ever made. The radicalism of

Nietzsche's thinking is largely due to the way this doctrine places in question most of our accepted ideas and assumptions about knowledge, life and the world. But it also implies a positive task. As Volker Gerhardt puts it, 'The fundamental philosophical question for Nietzsche is the question about the possibility of thinking *becoming*.'[4] One aspect of this involves seeing how becoming is related to thinking in its familiar form. Even if that is a false conception of the world, it has to be accounted for. What makes it possible in the first place? The question is a challenge to find out where ideas of time – that is, of duration and succession, of past, present and future – come from.

Nietzsche has his own answers to these questions, but he is also a philosopher of time in that much of his thinking about human life centres on this theme. Most notably, his critique of moral concepts such as responsibility, guilt and punishment involves tracing their source to the problematical human relation to past and future. In consequence, the mission of overcoming the prejudices of morality that he assigns to himself depends on the possibility of uncovering new forms of temporality. With this in mind, Nietzsche uses words such as 'moment' and 'hour' in ways that challenge familiar conceptions of time, and often draws on metaphorical modes of expression to achieve the same purpose. These different ways of grasping past, present and future have a direct bearing on his understanding of his own thought. He identifies himself as an 'untimely' thinker, one whose attitudes and values do not go with the contemporary social and intellectual world but instead run ahead to a more or less distant future. How such an anomaly can occur is, in fact, one of the issues that his thinking about time promises to address.

Although they are closely related, time and becoming are not the same thing. Since one of my main tasks will be to explore their interrelation, it will help to give a brief first sketch. For Nietzsche, becoming is a fact but time is an interpretation. The conceptual structures of temporality – that is, the categories of earlier and later, and of past, present and future – involve an elaboration of what is given in experience. They are not to be taken simply as descriptions of reality, or as generalizations from perception and observation. Rather, they have the properties of what Nietzsche calls 'perspectives'. This does not mean that we cannot compare them, or find ways of deciding between their claims. Every interpretation of becoming takes place within a context and cannot be understood or assessed in its own terms alone by means of a logical or philosophical analysis. We can call this context a 'form of life'. In Nietzsche's view, the same is true of all our concepts. Every perspective on reality arises out of a certain form of life and expresses the interpretations and, in particular, the valuations that are its conditions of existence. For example, a strong and active metabolism will inevitably interact with the world quite differently from a weak and reactive organization. Its attitude towards the future and past will express a sense of power and freedom and will support its drive to overcome obstacles, both outside and within itself. Nietzsche's genealogy of morality is based on these ideas, but he also applies them to knowledge in general, including concepts of time, space and causality.

Forms of life are complex and elusive in their various aspects, and Nietzsche often relies on figurative ways of describing them. What he says about time and becoming is nearly always bound up with metaphor. His spokesman Zarathustra draws attention to this: 'But it is of time and becoming that the best parables (*Gleichnisse*) should speak.'[5] An understanding of his work will depend on a deciphering of these metaphors and metonymies. That is always a contestable exercise, and others have come to conclusions that differ from mine. Yet we can often find reasons for deciding between these claims, for Nietzsche does not always speak in figurative language. The narratives and discourses of *Thus Spoke Zarathustra* are linked with what he says elsewhere in more familiar terms – that is, in the vocabularies of philosophy and of everyday life. For that reason, my treatment will move back and forth between different idioms, enabling them to interact in fruitful ways.

No interpretation is ever the only possible one, and so to speak of interpretation is to recognize alternatives. Nietzsche does not assume that there is only one way of making sense of becoming. As will be seen, he presents a number of distinct models of temporality, contrasted in various ways. This is disconcerting for anyone who wants to know which version is the 'right' one. Nietzsche would respond that there are only interpretations, none of which can claim the kind of truth that the question presupposes. With temporality, as elsewhere, his aim is not simply to reconstruct our familiar understanding of the world from a starting point of immediate experience – a familiar philosophical programme in his time, especially with the positivist school – but rather to open up further possibilities.

Some of these ideas were in the air when Nietzsche was writing. Philosophers had suggested that other sentient beings might experience space and time in ways quite different from human beings. Nietzsche speculates that beings with other conditions of existence could perceive space as having four dimensions.[6] Similarly, he wants to get away from thinking of time as a unique and necessary form of experience, the doctrine of the Kantian tradition. Yet to imagine that time is gone, says Zarathustra, is a kind of madness.[7] For it is a denial of the undeniable – that is, of the fact of becoming. The problems posed by time may seem insoluble, but we cannot escape the need to coming to terms with becoming, and this inevitably requires some perspectival interpretation, that is, some conception of time.

A Sign Language of Becoming

While Nietzsche is dogmatic about the reality of becoming, he is equally dogmatic on another point: the human mind is incapable of grasping becoming.

'*Knowledge* and *becoming* exclude each other', he writes.[8] One may wonder how Nietzsche can be so sure about this proposition. The justification that he most often suggests is an indirect one: a critique of our conception of change shows it to be a falsification of genuine becoming. This is certainly evidence of a circumstantial kind – why would we rely on that if an accurate version were available? – but it does not explain the necessity of falsification. In various places Nietzsche indicates sympathy with a biological epistemology in which sentience and knowledge are both explained by their contribution to the organism's life-process. This could be taken in a Darwinian sense, as providing an advantage in the struggle for existence – in other words, as contributing to increased fitness for survival and reproduction – or else according to the alternative model proposed by Nietzsche in his later period, centring on the concept of the will to power. Either way, the logic of the theory is that of natural selection.[9]

What is the most basic interpretation of reality that serves the purposes of life? Nietzsche writes, 'We cannot imagine *becoming* other than as the transition from one persisting "dead" state to another persisting "dead" state.'[10] Our cognitive bias is towards states of affairs that are stable rather than changing, and that in turn implies a belief in distinct substances or 'things', if not permanent, then at least persisting as identical even if changes occur elsewhere. This is clearly a departure from absolute becoming. An image of constant flow is conveyed in the best-known saying of Heraclitus, that we cannot step into the same river twice, since fresh waters are always flowing upon us.[11] Human thinking seems always to negate this continuity, only to face the task of dealing with experiences of alteration, coming to be and passing away. Nietzsche explains that these are grasped by contrasting the states of affairs coming before and after. He argues that the transition itself is *not* described directly, but left to be inferred from a succession of states that, taken by themselves, represent being rather than becoming. His criticism is that change is not captured in this formulation but, on the contrary, excluded in principle. In raising that suspicion, Nietzsche assumes and appeals to an implicit sense of the real flow of becoming, which shows it to be irreducible to other concepts, even the simple relation of before and after. Without such an awareness, we would never question a reduction of becoming to succession, or perhaps have needed it in the first place.

It is this introduction of discontinuities or 'gaps' between things that gives rise to a concept of time as something in its own right, and one can see why from the account already given. The assumption of persisting things entails a distinction between states of affairs and the locations that they occupy, since the same thing is posited as existing at different times. This is, one can say, a separation of the form and content of becoming. Belief in identical, unchanging things thus means that places in time, usually called 'moments', are distinguished from the objects, events or states of affairs that are located there. So it is that becoming turns into time, a conceptual adaptation so successful in terms of the evolutionary epistemology to which Nietzsche is drawn that the outcome seems never to have itself come about, but rather to possess a necessity that is independent of experience.

Nietzsche often describes the interpretative element in experience as involving what he calls a 'sign language' or 'semiotic'[12] designed to convey what cannot be formulated directly in a vocabulary that the mind can grasp. Sign languages, as he understands them, take a particular sense and adopt a terminology that refers to that sense. Our most basic way of representing becoming or change is as *motion*.[13] This familiar, seemingly literal concept translates diverse processes into a common language, in this case of vision – 'Movement is a symbolism for the eye', Nietzsche writes [14] – although touch presumably also plays a part. Motion involves concepts of the *thing* that moves, the *space* across which it travels and the *time* it takes to go from one place to another. Even if the world is understood as consisting of forces rather than material objects, those forces need to be imagined as things in order to appear in the mirror of our consciousness. Hence, the sign language of motion is still the standard means for representing change or becoming in human thinking.

What is wrong with this? one may ask. Nietzsche has a number of answers. One is that it is a rationalization for deficiencies in our powers of perception. He writes, 'We would not speak of time and know nothing of motion if we did not in a coarse fashion believe that we see things "at rest" beside things in motion.'[15] He calls this attitude 'coarse' because it displays an inability to perceive more than a small fraction of the changes actually occurring. This fact is continually confirmed by improved means of observation such as the microscope. Its consequence is that all the details that we overlook are assumed not to exist. Hence, Nietzsche concludes, we introduce 'lines' and 'surfaces' – that is, discontinuities – into the phenomena of experience. We separate rest and motion, although in reality there are only variations in the tempo of becoming, and those are differences in degree, rather than kind. There are also long-term changes on a scale beyond our reckoning. Again, Nietzsche appeals to recent discoveries in natural science: not just of organic evolution, but also of geological changes occurring over thousands and millions of years, requiring a new understanding of apparently permanent features of any landscape as phases in processes that are still continuing. With this in mind, he speculates that even the apparently eternal properties of chemical elements may just be very long-term patterns of relative stability. [16]

From time to time Nietzsche speculates that the cause of these errors lies in language and its determining influence on human knowledge. As one of his most quoted aphorisms puts it, 'Every word is a prejudice.'[17] Reification is built into the nature of language, or at least of the languages familiar to most of us. The Indo-European family of languages, he thinks, commits its users to a particular conceptualization of the world.[18] It designates objects as separate from one another and as enduring with identity. It requires us to postulate a subject for every process, even a very momentary one. We say that lightning flashes, even though there is nothing existing before and after the flashing that carries out some action.[19] Are there other languages that provide different and possibly more adequate representations of becoming? Nietzsche suggests that the Ural-Altaic languages (which include Finnish and Hungarian as well as various Asian languages) 'where the concept of the subject is least developed' may provide the best available alternative.[20] But he adds a qualification: if this did turn out to be the case, the real reason would lie in differences in '*physiological* valuations and racial conditions'.

Thus, despite his emphasis on the importance of language, Nietzsche comes down in the end in favour of a physiological explanation. Languages correspond to the forms of life of the peoples who use them, and especially to the valuations that are embodied in their conditions of existence. These are what enable them to survive and thrive, even to expand and conquer other peoples. A similar point applies to the perceptual apparatus that omits a great deal of the events around us and so produces a simplification of our environment. If we were aware of every detail of reality, we could lose the capacity for ready decision that is needed for fitness. Ignorance is an advantage, at least up to a point. Hence, while most human concepts may be falsifications, that is not to say they are arbitrary inventions. The errors of the human scale are not incidental to knowledge, but constitute its necessary basis.

Still, these concepts are certainly false in Nietzsche's judgement. We do not even need empirical evidence to prove this, because their inadequacies become evident on close examination. 'Our sensations of space and time are false,' he writes, 'for when tested for consistency (*consequent geprüft*) they lead to logical contradictions.'[21] We can locate the reasons for this failure in their origins. On his account, the familiar concept of change or becoming is formed by a process that has several stages. The first is the replacement of pure becoming with a postulation of enduring things that maintain their identity over time. The second is an attempt to recover becoming, given a recognition that change does occur and needs to be conceptualized in some fashion. The negation of becoming has to be negated in its turn, and that is where the logical problems begin. The reconstituted concept reveals its inadequacies when subjected to tests like those of Zeno's paradoxes, which expose the failure of its attempts to reconcile continuity and discontinuity, infinitude and discreteness.

In his Basel lectures on early Greek philosophy, Nietzsche presents the paradoxes of motion as directed against any belief in the 'absolute reality' of space and time. Zeno, he says, concluded that space and time possess no reality at all. However, a later 'profound critique of the intellect' has opened up a third possibility: that they are empirical realities for us, but not real in themselves. This is Kant's transcendental idealism and its solution to the antinomies of pure reason, but it comes through the neo-Kantian F. A. Lange, as Nietzsche's remark that 'we are required by our organisation to grasp everything under the form of space and time' indicates.[22] The paradoxes seem to be aimed against attempts to combine motion and rest in a single picture of reality. For example: the moving arrow is located in a particular place at every moment, but its motion cannot be equated with any totality of these states of rest. However, what Nietzsche suggests is that the concepts under attack are those of being and infinitude. Both are needed for space and time and motion, and yet they are incompatible, since Zeno shows that motion is impossible if the moving object has to traverse an infinite number of positions.

Alongside his physiological approach to the origin of categories of time and space, Nietzsche has another account using the language of drives and passions, pleasure and pain. In a note from the time of *Thus Spoke Zarathustra* he writes, '"Being" as the invention of the person suffering from becoming.'[23] This opens a renewed engagement with the themes of his earliest book *The Birth of Tragedy*. 'Becoming' there means a Dionysian world of continual and never-ending creation and destruction, of conflict and contradiction, which for sentient beings must be a realm of suffering. Standing in contrast is the Apollonian world of classical beauty, created by the imagination 'as a redemption from becoming'. In this setting belief in being appears as a reaction against the suffering that is bound up with becoming. Conversely, any affirmation of becoming must necessarily be an affirmation of suffering.

In this confrontational experience, Nietzsche suggests, we see the origin of the moral interpretation of the world. Coming to be and passing away are taken to be a punishment for wrongdoing, and so are both explained and justified. He attributes this view to Anaximander in his lectures, although in explaining the name of Zarathustra later on he suggests that the ancient Persian prophet of that name, thought to have lived some centuries before the earliest Greek thinkers, is the originator of the moral view of the world.[24] The moral interpretation of reality is not about beautiful illusion. Rather, it wants to relieve its own suffering by inflicting pain on others, in the belief that this will act upon and negate the source of its suffering, which it identifies as their past willing. What Nietzsche calls 'the spirit of revenge' is the attempt to rationalize and justify this project by passing off revenge as impersonal, objective 'punishment'. Further, the new focus on thinking and judging gives vengefulness a much wider application: it is the motivation behind a complete conceptualization of the world. The distinctions between subject and predicate, substance and attribute, agent and action, have a hidden agenda. They support a moral interpretation of the world by providing the basis for judgements of responsibility and guilt. Causality is supposed to explain what happens: in fact, Nietzsche argues, it serves primarily to assign responsibility and blame for our discontents and suffering. If agents are distinct from their actions, then it is possible to imagine them as carrying out some other action in the same circumstance, that is, as exercising freedom of the will. The concept of personal identity is also crucial for practices of blame and punishment, which require that the person at hand should the same as the one whose past action was the cause of present suffering. Its real function is thus to rationalize the moralistic standpoint which in turn is driven by motives of resentment and vengefulness. To trace the genealogy of these seemingly explanatory concepts through an account of their origins is to show how little they can be taken at face value.

The form of life that ideas represent is the ultimate criterion for Nietzsche, despite his recognition of a distinct logical critique. The dominant tradition of human culture for thousands of years is, he thinks, one driven by reactive forces which correspond to a weak and declining rather than a strong and creative form of life. It is because of this that we are motivated to devalue becoming and to privilege a world of being as the source of all value and meaning. Compared to this higher realm, the world of experience is inevitably judged as inferior. The same thing occurs with empirical experience, where what is permanent is counted as the ultimate reality to which all genuine knowledge must refer. As a scientific programme, this approach is certainly successful. The tradition of atomism, together with the modern laws of conservation of matter and force, has advanced the human capacity to grasp and control the world. Yet in the end it is all a fiction in Nietzsche's view. The will to truth impels us to uncover what lies beneath the elaborate construction of human knowledge – that is, to rediscover universal becoming, and to admit that the goal of transcending the world of appearance is unattainable.

The Decline and Fall of Being

Although these errors have dominated Western thought since ancient Greece, it is only the cultural changes of modernity that provide an opportunity to eliminate them within philosophy. After two thousand years of intellectual hegemony, Nietzsche claims, the doctrine of being has collapsed under its own pressures. The concept's trajectory is described in *Twilight of the Idols* under the heading 'How the "True World" Finally Became a Fable'. As this title suggests, Nietzsche sees the process as one of steady decline: from the dogmatic certainty of the original doctrine to weakened and qualified versions that lose all content. Within philosophy, this is seen in the retreat from Plato's doctrine of forms to the inaccessible thing-in-itself of Kant. The postulation of a 'true world' over and above that of experience began with a conviction that the sage could attain and belong to this reality. Later it became only a promise or aspiration, unattainable in the present life but held out as a hope of future reward for the virtuous person. Later still, the concept turned into an authority defining our moral obligations, or a guiding ideal, still there but no longer claiming to exert any direct influence over human life.

The ultimate form of belief in another world, Nietzsche says, is a complete separation of the true world from the world of appearance. Kant had prepared the way by making the world of being not only unknown but unknowable, but it was several writers of Nietzsche's own time who completed the process. The neo-Kantians proposed to eliminate this element from the critical philosophy altogether – in the name of Kant, or at least the Kant of the 'Transcendental Aesthetic'.[25] But Nietzsche had a higher regard for some recent thinkers who were prepared to accept a bold dismissal of everyday experience as a whole – including our immediate acquaintance with our own bodies – in the name of an epistemic ideal, submitting themselves to a pure logical criterion of truth.

A Parmenidean insistence on the absolute status of identity was championed by the Russian-born philosopher A. Spir in several books read by Nietzsche. It was not the doctrine of being that Nietzsche admired in Spir, but his integrity in following a single principle to its final conclusion. In the end, Spir wrote, there could be no relation at all between the unconditioned and the conditioned. Hence, the world of appearance could not be explained in terms of anything outside itself. Nietzsche draws his own conclusion: 'The "true" world – an idea which is no longer good for anything, not even obligating – an idea which has become useless and superfluous – *consequently* a refuted idea: let us abolish it!'[26] He calls for an affirmation of both appearance and becoming, simply because they are all that remains as the outcome of a long historical development. Tracing the imagined course of a new day from dawn to bright morning, Nietzsche concludes by announcing: 'Noon: moment of the shortest shadow: end of the longest error; high point of humanity; INCIPIT ZARATHUSTRA.' In other words, the stage is set for the entrance of his own thinking.

Nietzsche felt that the spirit of the nineteenth century favoured the thesis that everything is to be understood by how it has come about. Hegel's system had turned philosophy itself into a historical phenomenon, and Darwin could not have existed without Hegel. Taken by itself, this is a fanciful comment: Darwin knew nothing of Hegel, and was sceptical of any contribution to scientific knowledge not founded on empirical observation. Within a broader context, Nietzsche's remark makes some sense. The outlook that he adopted in his philosophical partnership with the Darwinist moral philosopher Paul Rée was what he and others called the 'historical' outlook. Its premise is that everything in experience *has become.* 'Lack of historical sense is the family failing of all philosophers', he writes at the start of *Human, All-Too-Human.*[27] Most of them have supposed that the human mind as it now exists has no origin that explains its nature. 'They will not learn that man has become, that the faculty of cognition has become.' Moreover, Nietzsche adds, they do not appreciate that the crucial stages of development took place before recorded history. Geology had opened up a vista of past time extending far beyond human records in which questions about origins could find ample room for speculative answers. The Darwinian account of the evolution of species was just one part of this universal outlook which was already supported by other thinkers of the time such as Herbert Spencer.

There were still areas of knowledge in which being rather than becoming remained dominant. Physical science was one that interested Nietzsche greatly. His rejection of atomistic materialism, the scientific orthodoxy of the nineteenth century, was closely linked with his critique of any conception of time assuming identical, enduring substances. The mobile but unalterable atoms of natural science were a further example of the denial of change implicit in everyday experience. Nietzsche hoped that the dynamic physics of Boscovich could exert a countervailing influence against materialistic atomism. Yet even this alternative retained much of the atomistic prejudice: Boscovich's points of matter are still permanent identical entities, possessing inertia and impenetrability even if lacking the qualities of extended matter. But what further step could there be?

Nietzsche's answer is indicated in the brief outline of an entire theory of space, time and matter that he sketched out in an 1873 notebook.[28] Long overlooked, this text has attracted attention in recent years. Nietzsche sets out to devise a physical theory which will complete the process of replacing Parmenidean atoms with Heraclitean centres of force, purely momentary in their existence and yet capable of explaining the patterns of becoming. He calls them 'time-points' or 'time-atoms', by analogy with the indivisible particles of scientific atomism. They are the theme of a philosophical experiment, a theory of pure succession in which each moment is an absolute reality, and becoming is attributed to elements whose existence is purely momentary. Fragmentary though it is, the time-atom sketch points to an ambitious programme. As I will show, this is nothing less than a Heraclitean philosophy of nature, in which patterns of time figure as a theoretical basis for grasping empirical reality. The time-atom theory proposes a radical shift from 'laws of space' to 'laws of time' in order to reconstruct our knowledge of the world. It envisages a reduction of the regularities of natural phenomena to non-spatial patterns of 'motion in time'. Nietzsche's earlier studies of meter and rhythm gave him an understanding of time-relations as forming recurring and concurrent sequences. How the theory is related to his later thinking is not easy to say, though. It stands quite apart from his usual insistence on the absolute continuity of becoming, and yet in other ways it has much in common with later speculations. So, despite its untypical features, the theory is linked with both earlier and later writings. It takes up elements of the theory of rhythmic form, and it points ahead to renewed engagements with atomism, always a doctrine of special importance for Nietzsche. It would be wrong to think of the time-atom theory as a temporary false path in his development, as we shall see in Chapter 3.

Becoming and Time

Before considering what Nietzsche has to say about time, we need to explore his conception of becoming, and his reasons for endorsing a Heraclitean view of the world. The first two chapters of this book will address these issues. My starting point is Nietzsche's statement of his own commitment in the conflict between the doctrines of being and becoming. The world must always be in the process of changing, he argues. If it were capable of standing still even for the briefest moment, then a state of rest would have occurred at some past time. In that case, the course of becoming would have come to an end, since there would be no reason for it to begin again. The self-evident reality of becoming shows that this is untrue. Thus, we can demonstrate that the world will never reach a final state, despite the speculations of physical science on an eventual standstill of force.[29]

How is that impossibility to be explained, though? Nietzsche's answer is that continual change is an essential feature of reality, understood by him in terms of force: 'it cannot stand still'.[30] This claim needs further elaboration, and Chapter 2 sets out to find what aspects of Nietzsche's concept of reality – in whatever sense he can be said to have one – lead to the conclusion that Heraclitus was right in affirming absolute becoming. Another question is closely related: what aspects of reality enter into the interpretation of becoming as time and make it possible to conceptualize what is given in experience in this way? Nietzsche's answers are made harder to identify by his preoccupation with attacking the familiar concepts of succession and causality. His own alternative tends to emerge only as part of this criticism. However, given his affinity with the ancient Greek thinkers, it is not surprising that one of their main themes is prominent in his approach. The oldest philosophical accounts of becoming depict it as a continual process of conflict between opposite qualities: hot and cold, wet and dry, black and white, and so on. For Nietzsche, too, becoming involves conflict or, rather, a continually shifting pattern of conflict and concord. That is a theme which will run throughout this book, and the presence of tensions and contradictions will later be seen to play an important role in interpretations of becoming as time.

In *The Birth of Tragedy* conflict is the underlying reality of existence and the original experience out of which arose the Greek insistence on beautiful illusion, on a drawing of boundaries and establishment of forms. That has much to do with time, and it may be argued that time in its original appearance is an Apollonian form, the structuring of becoming according to a regular measure. 'Apollo is the god of rhythm', Nietzsche writes.[31] In the period of *The Birth of Tragedy* he understands rhythm as an aesthetic phenomenon – but it is precisely through such aesthetic phenomena that human beings have made their first attempts to come to terms with the ungraspable and deeply threatening world of becoming.

What is there about becoming that allows it to appear in both of these ways, as the basis not only of conflict and contradiction but also of stable, regular patterns? That variations of some kind occur in the flow of becoming is the first postulate. The atomistic scheme of qualitatively identical things moving and interacting in space cannot be assumed. Instead, the primordial variability of absolute becoming is a *fluctuation in intensity*, to borrow an expression of Pierre Klossowski, that occurs prior to any construction of extensive magnitude in space or time.[32] Once again, the problem is to make this accessible to human knowledge. A sign language of the senses is needed if variations in intensity are to be grasped by the mind. Given that the primary image of becoming is motion, the natural symbol for these fluctuations is what Nietzsche calls 'variations in tempo'.[33] These are in turn experienced as contrasts between rest and motion, or persistence and change, which provide the basis for a construction of becoming as succession and rhythm, and finally as time.

From Chapter 3 onward, conceptual models of time will be discussed. I make a distinction between two ways of considering time. One refers to what Plato calls the *parts* of time: days and nights, months and years. These are periods or intervals, originally defined by the regular motions of the heavenly bodies. Because each of these objects undergoes recurring cycles, it provides a *measure* for time – that is, a standard for determining how long or short any other period of time is. The other model of time involves what Plato calls the *forms* (*eide*) of time: past, present and future.[34] For him, these are bound up with change and becoming. They are sharply contrasted with the timeless reality of which the world of becoming is only a 'moving image'. We use *tensed* language, 'was', 'is' and 'will be', in referring to the forms of time, but only 'is', Plato explains, can properly be applied to what is eternal and unchanging.[35]

Both of these models will be prominent through the following discussion. As we will see, they are not the only ways in which Nietzsche approaches the nature of time. But of the ideas that he draws on for that purpose, they are the most familiar at first sight. Not only are they in common usage, but they have been extensively explored by philosophers. Writers disagree over whether both of these systems are needed for thinking about time or, to use a grammatical terminology, whether tensed language is reducible to untensed language and so eliminable. These debates are not just about words: the reality of becoming is at issue. If absolute becoming seems difficult to grasp, it should be appreciated that so is becoming by itself. Philosophy has struggled to formulate it as a concept in the face of Zeno's paradoxes and later puzzles. One solution is to give up the attempt and provide a substitute for becoming which avoids such issues. After all, it is argued, we can say most of what we want about time without using the language of becoming. Events will still occur at different moments, occur before and after one another, and take up different amounts of time, and propositions about present, past and future can be restated in that more limited vocabulary.

A reply to this position is that it leaves out something essential to time, something that we encounter in immediate experience, or so it is claimed: the reality of *becoming*. Without that, one cannot claim to be truly speaking of time. But, a sceptic will ask, is there really such an experience? What if 'becoming' is just another word for succession? That can be characterized as one event's occurring later than another, without any mention of past, present and future. Now, assertions about immediate experience are always open to criticism, especially where certainty is involved. Nietzsche is highly suspicious of claims for self-evidence and ready to deny their role in supporting philosophical positions, yet he makes an exception in this case. He maintains that our own thinking provides a 'single certainty', the reality of becoming, even if other claims made about it – concerning an identical self, for example – are based on unjustified interpretations of what is given.

More recent writers have been less willing to rely on the evidence of an immediate intuition to validate the notion of becoming and the distinctions between past, present and future. Defenders of the 'tensed' view look for more subtle strategies. They draw attention to aspects of human experience involving feelings such as hope, fear, regret and relief, in which the past or future status of the intended object is essential to the feeling itself. Their case is supported by concepts that express human attitudes towards past or future and that cannot be accounted for without the use of irreducibly tensed language. Much of this philosophical debate is relevant to Nietzsche's treatment of time. As a defender of becoming, his allegiance is to the side that insists on the necessity of referring to past, present and future in addressing the reality of time. This does not mean that Nietzsche has nothing to say about the 'parts of time'. In fact, he is often concerned with these themes, for example, with the question of a quantitative 'measure' of time that provides a way of dividing time into definite periods. But the forms of time are central to his thinking.

We could carry out an investigation into Nietzsche's theory of time using a philosophical vocabulary, but it is characteristic of him that such abstract concepts are supplemented with figurative images whose role is crucially important. In the central chapters of *Thus Spoke Zarathustra* he employs a scheme of metaphor to express several modes of temporality. The details are not always spelt out, and the task of unpacking each version places a great deal of responsibility on readers. Part Three of *Thus Spoke Zarathustra* is the key text for our purpose, because it explores a sequence of metaphors which carry a range of meanings, involving forms of life and the temporalities that correspond to them.

These metaphors are images of *ways* on which movement or travel occurs. In the chapter 'On the Vision and Riddle', there is first a mountain path and then a gateway at which two lanes come together. The previous chapter, 'The Wanderer', speaks of a 'way of greatness'. Each of these four ways (counting the gateway and the lanes separately) involves a situation in which particular forces enter into relations, sometimes of tension and conflict. Each is an interpretation of a reality – the fact of becoming – which the mind can grasp only through imposing conceptual forms that serve its own conditions of existence. But these are different for different forms of life. Hence, to these ways correspond kinds of movement, such as walking, running, climbing, dancing and flying, and I will argue for the inclusion of standing in this list to make it complete. In each instance, the metaphor includes not only what the motion is, but also some of its features: for instance, what direction or directions it proceeds in, how far it extends, whether many or few engage in it and what kind of interaction of forces is experienced.

All this provides Nietzsche with a complex metaphorical vocabulary, well suited for types of temporality that are otherwise quite elusive. The imagery of Zarathustra's four ways performs this function in considerable detail. The temporal meaning of one image is explicitly stated: we are told that the gateway standing between two long lanes has a name written above it: 'Moment' (*Augenblick*). This implies that the gateway stands for the present moment, and the lanes for past and future time. For other images, a harder task of interpretation is at hand. It is not evident that when Zarathustra describes a mountain path the image he draws has some temporal connotation. However, I will be arguing that this way is his symbol for a particular model of becoming, one that represents time as consisting of finite periods which can be counted or measured.

A comprehensive survey of the kinds of motion that Nietzsche has in mind is provided when he writes, 'He who would learn to fly one day must first learn to stand and walk and run and climb and dance: one cannot fly into flying.'[36] A key feature of this list is its clearly defined order. As we move through the forms of motion, a progressive increase in freedom and self-directedness is evident. Each mode in turn displays a greater scope for activity and, accordingly, the corresponding locations range from a single place for standing, through to an unbounded expanse of space in the case of flying. In fact, the ordering signals what Nietzsche calls an 'order of rank'. That implies that it is connected with broader ideas about personality, power and freedom – that is, with leading themes in his thinking as a whole.

In Chapter 4 I begin this discussion, starting with the dramatic narrative of the chapter 'On the Vision and Riddle'. To an audience of sailors, Zarathustra describes events which, he says, have occurred recently. The scene is set in a mountainous landscape at dusk. As the episode begins, Zarathustra is climbing a steep and dangerous path, weighed down by the burden of a strange dwarfish being: this is the 'spirit of gravity', who draws him downward towards the abyss. The mountain pathway recalls images of steps and ladders elsewhere in *Thus Spoke Zarathustra*. Movement occurs there in discrete steps or stages, and so provides a picture of time consisting of finite intervals which are countable or measurable according to some standard. What is involved in the activity of climbing adds further details to this concept. In particular, a conflict of forces cannot be overlooked. Climbing entails confronting and overcoming resistance with every step. Looking more closely at Nietzsche's images, we see that there is another tension, between opposing impulses within the individual person. It is because conflict is intrinsic to becoming that tension and contradiction figure within the construction not only of parts of time (such as successive days) but also, as we shall see in Chapter 5, of the forms of time, past and future.

Zarathustra halts to confront his hostile companion and finds himself at a gateway from which extend two long lanes (*Gassen*), one stretching forward and the other backward. The two lanes come together in this gateway, and they are said to contradict each other and collide 'head to head' there. Now Zarathustra poses several challenges to the dwarf, who fails to reply successfully and then disappears. His first question is: do these lanes contradict each other eternally, assuming that they do so at the gateway 'Moment'? In response, the dwarf says that 'time itself is a circle'. Not only does this not directly answer the question asked, but it is itself a puzzling assertion. Commentators have disagreed on whether Zarathustra accepts or repudiates his response, and even on whether the dwarf means what he says. Moreover, it has been suggested that the concept of a circular time – that is, a time closed in form – is somehow incoherent. Nietzsche's text does not resolve these issues: the initial interchange ends inconclusively. It is immediately followed by another challenge, different in both style and content, not about the lanes but about those who travel on them. Zarathustra asks, 'Must not whatever *can* run have run on this lane before?' Presumably so, one gathers – but that speculation leads on to a new riddle: must not everything also return again eternally, out on these distant lanes? In that case, the present moment would also recur, along with all that it contains, the encounter and its participants. Zarathustra raises the idea that whatever occurs does so not just once, but again and again – in fact, infinitely many times. This is a bold assertion, yet not an arbitrary one. It is introduced through a sequence of questions which, in fact, amount to a line of reasoning with this proposition as its conclusion. Given certain assumptions, it must be supposed that whatever occurs does again and again, in the same order. Of course, the truth of these premises is another matter. Still, the common notion that an argument for eternal recurrence is found only in Nietzsche's unpublished notes is simply wrong.

Thematizing the eternal return fully would mean addressing questions that lie outside the scope of this book. Its bearing on human existence is what dominates an extensive secondary literature. These readings tend to suffer from a failure to locate the thought within Nietzsche's ideas on time and becoming. What is meant by the recurrence of the present state of affairs, as I will argue, is not its occurring at earlier and later times, but rather its occurring in the past and in the future. That may seem to amount to the same thing, but the differences emerge in considering our attitudes and feelings. As I have noted, the past and future are what we care about: hope, fear, regret and other emotions are directed towards events characterized in those terms. The feelings associated with eternal recurrence may not be identifiable with these familiar ones in any straightforward way, given the epistemic uniqueness of the concept, but they are still bound up with notions of past and future, and this comes through in the way that Nietzsche presents the doctrine, especially in *Thus Spoke Zarathustra*. The discussions in Chapters 6 and 7 will set out to throw further light on this aspect of the thought.

Chapter 8 looks into an alternative to the models of time discussed so far. It is the idea of a temporality which is separate from the universal time of the world and can be said to form a time in its own right. My point of departure here is the fourth of Zarathustra's 'ways', referred to in cryptic fashion as the 'way of greatness' (*Weg der Grösse*). This is the most difficult of Zarathustra's ways to grasp, even as a picture. Nevertheless, there is a time paradigm that corresponds to it. The word used for this in *Thus Spoke Zarathustra* is 'hour'. The hour is not a part of time, but is a self-contained totality. Further, what is unusual about the hour is that the time and its content are not separate. In this sense, it is almost a return to absolute becoming, which also lacks any distinction between form and content. The difference is that in the hour, the forms of time are not negated, but rather incorporated. The conflict between them which gave rise to their confrontation at the gateway 'Moment' is reconciled. The 'flowing apart'[37] that Nietzsche saw as the impression made by the original Heraclitean insight is replaced by a gathering and concentration. In the hour, past and future are able to dwell together.

It would be an error to suppose that this amounts to a transcending of time, a new version of the traditional metaphysical escape from becoming. The conflicts that characterize the other 'ways' are suspended in the hour, but not eliminated once and for all. The hour has an ending as well as a beginning, and when it ends, 'the hand moves forward, the tragedy *begins*'.[38] The highest realization of becoming that Nietzsche envisages is not this reconciliation of past and future by itself, but rather an affirmation of the continual and never-ending alternation of conflict and harmony, moving 'out of the play of contradictions into the joy of concord'[39] only to recommence the whole process. Hence, it does not wish the hour to stay but instead says to it, 'Go, but return.' The Dionysian is an affirmation of becoming in its aspect of creation and destruction. 'Becoming, felt and interpreted from within, would be the continual creation of an unsatisfied, over-rich, infinitely tensed and constrained being, a god, who overcomes the pain of being only through continual transformation and change.'[40] Such an affirmation of becoming is joy in building and destroying even the most beautiful illusions. This book's interpretation of Nietzsche's thinking about time and becoming concludes with an attempt to address this conception. Here too we find him postulating a form of life, one for which only the symbol of the god is adequate. Whereas Cratylus, the most faithful disciple of Heraclitus, was unable to speak but only moved his finger, the Nietzschean god has no need to use words, but dances with an expressiveness that says all that in the end needs to be said about becoming.

Chapter 1

Absolute Becoming

In his incomplete survey of early Greek thought, *Philosophy in the Tragic Age of the Greeks*, Nietzsche describes an imaginary meeting of the ancient thinkers Parmenides and Heraclitus. This 'unhappy encounter', as he calls it, occurs on the path that leads down into 'the abyss of all things'.[1] The two pass by each other only to exchange hostile and dismissive words. Their meeting is unfortunate because there can be no communication, let alone dialogue, between thinkers who have no common language, but only a hostile confrontation.

How does this strange impasse come about? It is the human counterpart of a conflict which, for Nietzsche, defines the whole course of Western thought. His account centres on the concepts of being and becoming, and of reality and appearance. Philosophy begins with attempts to provide a general theory of the natural world, to make sense of its continual changes and transformations, some regular but others unpredictable and puzzling. The earliest answer came from Anaximander, reported by the Aristotelian commentator Simplicius as asserting that the source of existing things is the 'indefinite' (*apeiron*), and that this is also where they return on ceasing to be 'according to necessity, for they pay penalty and retribution to each other for their injustice according to the assessment of Time'.[2] Nietzsche quotes this text (although without the words 'to each other') and comments: 'All becoming is an emancipation from eternal being: hence an injustice, and hence subject to the punishment of perishing.'[3] On his reading, the 'injustice' is the separation of particular things from the unity of the unlimited. It is the assertion of individuality that is their offence, and the penalty is an enforced return to the undifferentiated source. In the longer version, however, Anaximander is identifying the injustice as committed by existing things against one another. Given the mutual opposition and conflict of influences such as heat and cold, each thing arises only in so far as another has ceased to be, and must balance its account by giving place to some new thing in turn.

On either reading, the moral interpretation of reality is asserted. If passing away is the punishment that things incur simply by virtue of coming to be, then the moral categories of guilt and retribution are not just human conventions, but extend through the world as a whole. There is no escape from the inevitability of injustice and condemnation. It is this thought, with its gloomy implications for human life, that makes Anaximander 'the first pessimist philosopher'.[4] Nietzsche sees Anaximander as the key figure for all subsequent philosophy, in that he introduces the dichotomy between two realms, of being and becoming, and poses the problem of the relation between them. In his system, the particular things that come to be and pass away arise from the original source that is designated only by a negative expression, the 'unlimited'. All such things are subject to change and passing away. For this same reason, Nietzsche says, the indefinite must be timeless: not because it is infinite, but because it has no properties that could be negated by opposed properties, causing it to pass away.[5] It is an equivalent of the unknowable 'thing in itself' of Kant's philosophy. Correlated with the contrast between being and becoming is that between reality and appearance. Again, the moral aspect of the doctrine imposes a sharp contrast in valuation on the distinction. Appearance declares its lack of true worth by continually negating its own existence, whereas genuine reality is exempt from the judgement of time, and so is permanent.

The importance of Heraclitus lies in his vindication of both becoming and appearance. In *Philosophy in the Tragic Age of the Greeks* Nietzsche introduces Heraclitus by contrast with his predecessor Anaximander. He imagines Heraclitus as contemplating the 'eternal wavebeat and rhythm of things'[6] and finding not injustice but only order and lawfulness in the world of becoming. Becoming is 'innocent' just because it owes nothing to an underlying reality from which it has separated itself and, in doing so, committed an offence deserving punishment. Hence, Heraclitus denies a dualism of physical and metaphysical worlds and takes a bold further step by denying being altogether. 'I see nothing other than becoming', he declares in Nietzsche's narrative.

Becoming as Heraclitus characterizes it is driven by the continual conflict between opposites such as light and dark, bitter and sweet, which are never separated but always entangled with one another, in Nietzsche's words, 'like wrestlers of whom now one, now the other gains the upper hand'.[7] What we perceive as definite qualities are these temporary phases in the never-ending contest. 'Everything happens in accordance with this strife, and it is just in this strife that eternal justice is revealed.' This conception is crucial to Nietzsche's reading of Heraclitus. If becoming is to be free from moral judgement, it cannot be driven by motives that seek either to achieve some goal or to avoid something, or even by drives such as love and hate. It must be a *playful* activity. Nietzsche often refers to an expression attributed to Heraclitus, responding to the question 'What is time?': 'A child playing, moving counters, gathering and scattering.'[8] Hence, his own description of the world of becoming as 'the endless game of foolishness that the great child time plays before us and with us'.[9] For Anaximander, things pass away because they deserve to do so, but this assumes a higher authority that passes judgement and enforces punishment. For Heraclitus, in contrast, there can be no judge superior to the contestants themselves. The disagreement is thus between a moral and an aesthetic view of the world: one imposing judgement and punishment, and the other affirming 'the innocence of becoming'.

This gives us one of the participants in the 'unfortunate encounter'. What about the other? Parmenides is presented as the opposing counterpart (*Gegenbild*) of Heraclitus. Like Heraclitus, he rejects the separation of a world of being and one of becoming. But whereas Heraclitus relies on intuition for his philosophical insight, Parmenides appeals to pure logic. Like other scholars, Nietzsche is faced with the puzzle of reconciling the uncompromisingly radical first part of Parmenides' poem with the second part, which presents a philosophy of nature related to that of Anaximander. He solves the problem by claiming that these represent two phases in Parmenides' development. An earlier system involves a world of positive and negative qualities, brought together by the power of attraction. On Nietzsche's account, Parmenides abandoned this at some critical moment when he became convinced that the single principle of logic, the law of identity, disallows all negative qualities. What has being cannot have come to be, because in that case it would have merged from non-being, an impossible concept. The same argument eliminates passing away, plurality and movement. For Parmenides, then, becoming is an illusion of the senses. 'All sense perceptions, Parmenides claims, offer only deceptions; and their main deception is just that they pretend that the non-existent also exists, that becoming also has being.'[10] Reality must be without plurality, an indivisible oneness. Further, it must be without change – in fact, timeless. As Plato puts it, we cannot say that it was or will be, but only that it is.

The aftermath of the dispute between Heraclitus and Parmenides is a new theory which attempts to incorporate both being and becoming in a single system. It acknowledges the reality of change, yet insists that what is cannot come to be or pass away. The world must consist of things that have existed from eternity, and yet are subject to 'so-called becoming' in that they are capable of motion, and so continually alter in their relations to one another.[11] In the words of Anaxagoras, 'Nothing comes into being or perishes, but from things that are it is compounded and dissolved.'[12] In one sense this makes Parmenides the winner in his dispute with Heraclitus, however implausible his doctrine in its pure form. For the Eleatic philosophy gave rise to atomism, a programme of unique importance for scientific thought.[13] The atoms of materialism are not only changeless but indistinguishable, since they have no qualities of their own: those arise only through interaction with a perceiving subject. Subsequent natural philosophers explain becoming as a recombining of elements which do not come to be or pass away, but only enter into varying combinations or configurations. Is this thesis a denial of becoming? Certainly it acknowledges change, and yet the elements of reality remain eternally the same. 'We must never speak of becoming here', Nietzsche insists, explaining that this word should be reserved for genuine coming to be and passing away.[14]

By identifying the atoms of modern physical science with Parmenidean being, Nietzsche is able to argue for a link between ancient thought and his own preoccupations. In the same way, he presents the 'historical school' of the nineteenth century, with its insistence on understanding any phenomenon by investigating its origins, as the revival of a Heraclitean way of thinking.

What that means is that the dispute between these two thinkers is not yet over. As I will show in Chapter 3, Nietzsche's ambition is to confront physical theory on its own ground, and to reaffirm a Heraclitean science of nature.

The Reality of Becoming

What reason could we have for not accepting the Parmenidean denial of becoming, supported as it is by the power of logical reasoning? Nietzsche's answer appeals to immediate experience. In *Philosophy in the Tragic Age of the Greeks* he presents Parmenides as praying to the gods to grant him just one certainty. The request is granted with the insight that 'what is, is, and what is not, is not'.[15] Nietzsche is later to make his own claim: 'This is the sole certainty we have in our hands to serve as a corrective to a great host of world hypotheses possible in themselves.'[16] He means the fact of becoming, which cannot be denied without evident absurdity. It is the only certainty that he concedes, and yet it supports a dramatic sequence of consequences, just as the single premise of Parmenides does. The present state of things is one of ongoing change, not of rest: Nietzsche argues that this is confirmed by our immediate awareness of our own thinking. Usually he is sceptical about appeals to the facts of consciousness. Even to say 'I think', as Descartes does, is to build in metaphysical assumptions about subject and process, based on the atomistic prejudice and the linked assumption that every process is an activity dependent on but distinct from its agent. But in a minimal form, as an affirmation of becoming, the *cogito* argument is presumably harmless. It says nothing about a subject of change, and leaves open the option of an eliminative interpretation of the kind that Nietzsche favours.[17]

In *Philosophy in the Tragic Age of the Greeks*, Nietzsche uses this line of thought as, in his own words, an argument *ad hominem*. Its target is the Eleatic use of logical argumentation against the reality of becoming and multiplicity. He writes, 'If the thinking of reason in concepts is real, then multiplicity and motion must also have reality, for rational thinking is in motion, and is just a motion from concept to concept, that is, within a multiplicity of realities.'[18] However, he imagines Parmenides as offering a response used by Kant 'in a similar case, against the same objection'.

> I can indeed say that my representations follow one another; but this is only to say that we are conscious of them as in a time-sequence, that is, in conformity with the form of inner sense. Time is not, therefore, something in itself, nor is it an objective determination inherent in things.[19]

Kant is responding to what he says has been 'unanimously urged' as an objection to his account of time: that however sceptical we are about external objects, the changes in our representations are real, and so therefore is the time in which those alterations occur. He replies that he accepts the argument, as stated. Certainly time is empirically real – however, that does not make it an absolute reality. Now Kant makes a counter-attack. The critics go wrong in assuming that we can be mistaken about external things, but not about the objects of inner sense. In fact, both are just representations, and display features that are due to the subject, rather than present in the object in itself. These contributions are, first and foremost, the forms of intuition, time and space.

Spir disagrees strongly with Kant on this point, accusing him of inconsistency where time is concerned. For Spir, the doctrine of the ideality of time is an evident absurdity. His criticism is entertaining, if not altogether fair to Kant's position. 'Caesar and Socrates, according to his hypothesis, are not really dead. They are just as alive as they were two thousand years ago, and merely appear to be dead owing to an arrangement of my "inner sense".'[20] Kant's 'real' view, according to Spir, is that succession can be known only in relation to the presence of some identical thing – the argument of the 'Refutation of Idealism' in the Second Edition of the *Critique of Pure Reason*.[21] In that case, Spir argues, succession itself could not be perceived immediately, but only inferred. But this is just as unacceptable, since we do observe change occurring.

Spir's own view is that 'the reality of our ideas themselves cannot be doubted, and these are obviously themselves successive, incontestably following one another in actuality'.[22] Thus, change is just as much a reality as the rest of the empirical content of experience, from which it cannot be separated.

> The fact is that one cannot deny the reality of change (*Veränderung*) altogether. If it is thrown out the window, it will slip in again through the keyhole. One can say 'it merely seems to me that conditions and ideas change', – but then this semblance itself is something objectively at hand and within it, succession undoubtedly has objective reality; within it things actually follow one another.[23]

In quoting these words, Nietzsche adds a passage from a footnote in which Spir makes a stronger claim, attributing to Kant the insight that all our ideas (*Vorstellungen*) 'appear as they are'. If inner perception were not incorrigible, he argues, we could never have an epistemology or a transcendental philosophy possessing objective validity.

> Now it is without doubt that our ideas themselves appear as successive. If Kant did not wish to regard time as something actual, he was quite right. For time is a mere abstraction from real successions and cannot be imagined at all without these (i.e. as empty time). But disputing the reality of given successions themselves was an extraordinary undertaking.[24]

Bearing in mind that this is, after all, an argument *ad hominem*, Nietzsche explains that the reality of motion would still be considered in a Parmenidean fashion.

'Now if motion has being, then what is true of being in general and in all cases is true of motion: it is uncreated, eternal, indestructible, without increase or decrease.' And that, he says, brings us to Anaxagoras. 'But as far as Parmenides' main doctrine goes, he kept Anaxagoras in submission to it, as he did all subsequent philosophers and nature investigators. They all deny the possibility of coming to be and passing away, as ordinary people imagine it.'[25] So we are left with an inconclusive situation. Acknowledging that some things are changing at any given moment does not amount to asserting that everything is changing all the time. How does one get from here to an affirmation of universal becoming? To answer this question, we need to return to the Heraclitean principle of becoming and see how Nietzsche intends to use it.

The Doctrine of Absolute Becoming

Few later thinkers have endorsed the bold thesis of Heraclitus that 'everything flows' (*panta rei*). Some, like Plato, have postulated an eternal, changeless realm of being compared to which the world of experience, for which they concede that Heraclitus's claim may be true, is a derivative and lesser kind of reality. Others have accepted the common sense view that while change is common, some things are relatively stable and enduring, even if within limits, in that they come into being at one time and pass away at another. Some thinkers, especially those belonging to theistic traditions, have added that the world as a whole must have a beginning or an end, or even both. The doctrine of Heraclitus rules out all these conceptions. The flow of becoming that it affirms has neither beginning nor end, but proceeds from a past eternity to a future eternity. Within this course it has no pause or rest, not even the most momentary one. Thus, there is only becoming, and nothing else.

The objection made against this concept from Plato onward is that it makes true statements impossible – in Plato's words, 'But if nothing is at rest, every answer upon whatever subject is equally right.'[26] We cannot say what things are, he explains, or even what they become, for the 'thus' and 'not thus', referring as they do to properties and states, are no longer available. Aristotle's objection to absolute becoming is that by eliminating distinctions it produces an outbreak of contradictions: 'for it will be the same thing to be good and to be bad, and to be good and to be not good, and so the same thing will be good and not good, and man and horse'.[27] Whatever its claim to truth may be, the doctrine of becoming seems to make all organized knowledge of reality impossible. As Michael S. Green puts it, 'A world of absolute becoming is without substances and causal relations, within which things happen at no particular rate of time and things exist in no particular place.'[28] It is impossible to make meaningful statements because even the simplest concepts used for describing the world are inconsistent with the thesis that change is occurring everywhere and at every time.[29] Thus, we find ourselves entangled in contradictions or, what is just as bad, left without determinate statements. In short, absolute becoming is not a *useable* concept for human knowledge.

The doctrine of absolute becoming is plainly inconsistent with common sense, which takes for granted the presence of enduring things around us. Nor does it fit readily into a naturalistic scientific philosophy, whose concepts correspond to accessible aspects of experience. As the historian of materialism F. A. Lange noted, 'materialism trusts the senses'.[30] Even when it questions the reliability of everyday observation, the concepts that it uses as replacements are based on our experience of solidity and motion. The physical science of Nietzsche's day was particularly dominated by a bias towards the category of being. Despite the rise of the 'historical school', it remained unsympathetic to any reaffirmation of becoming. Material bodies were interpreted as composed of atoms which never change in themselves, but simply adopt new combinations and arrangements. Nietzsche sometimes speculates that the question is just one of scale: atoms or centres of force are assumed by scientific theory to be unchanging, but he suspects that they may just be very slow by human standards in their alterations.[31] But the challenge of atomism continues to preoccupy him.

Despite all these issues, Nietzsche tells us that Heraclitus is right when he says 'I see nothing but becoming.'[32] The reality of becoming is the most basic fact of all, encountered in our own thinking in a way than cannot be denied without evident absurdity. In fact, 'we ourselves are, indeed, nothing but that which at every moment we experience of this continued flowing'.[33] Thus, Nietzsche rejects Parmenides and aligns himself with Heraclitus, accepting that the category of being has no proper application. 'Heraclitus will remain eternally right with his assertion that being is an empty fiction.'[34] What this means is that there are no 'things' undergoing alteration.

> The tree is something *new* in every moment: the *form* is maintained by us, because we cannot perceive the finest absolute motion: we impose a *mathematical line* into absolute motion, in general we *introduce* lines and surfaces there on the basis of the intellect, which is the *error*: the assumption of sameness and persistence, since we can *see* only what lasts and *remember* only what is similar (the same). But in itself it is different; we need not carry our *skepsis* over into the essence of things.[35]

All the more or less permanent things we seem to encounter in our daily experience are not what we take them to be. Instead, becoming (*Werden*) is identified with occurrence (*Geschehen*) or process (*Prozess*)[36] and Nietzsche explains that a process is not to be separated 'from a something that is not process but enduring, substance, thing, body, soul, etc.'[37]

One important source of Nietzsche's thinking about absolute becoming is Spir, who follows J. F. Herbart in holding that in absolute becoming 'change (*Wechsel*) itself is to be seen as the quality of what is subject to it'.[38] What can be said about things whose only property is change? If they do come and go continually, that means that they can have no causes and obey no laws, because these assume the possibility of correlation. But, Spir asks, can the content and form of change be one and the same? It is hard to see how we can combine concepts whose meaning seems to depend on their contrast. Perhaps what is meant is that whatever the content may be is 'a mere accident', not worth special attention, and all that counts as real is the form of change.[39] Furthermore, we cannot imagine anything real arising out of nothing. That would imply that nothing alters, that it becomes something, which is absurd. In any case, Spir adds, we have no experience of 'nothing'. Everything in experience must have some cause: that is the only rational assumption. If we do not know what something arises from, it is better to say that than to use a misleading expression. Spir's conclusion is that absolute becoming cannot be conceptualized. He quotes Herbart's further explication, which spells out the concept of time that corresponds to absolute becoming:

> This means that, first of all, that it does not alter at one time and stay the same at another, but that change proceeds continually, *from the whole past into the whole future*, without beginning, without pause, without end. Further, that it proceeds *continuously with the same speed*; so that in the same periods of time the same amount of alteration is always completed. Finally that the *direction* of alteration is and always remains the same, so that going backwards and again forwards, the repetition of previous states, is completely excluded. –[40]

Spir has no quarrel with the first of these descriptions, but objects to the other two. How can we attribute a constant speed or constant direction to absolute becoming? Such notions imply that a succession can be grasped as having these features. But such a synthesis is impossible with things that appear only to disappear straight away, leaving no trace behind. They would have to possess some character over and above change itself for any correlation of their states to be possible.[41]

Many of these ideas reappear in Nietzsche's writing on becoming. In one late note he sets out the implications of the proposition that the world is one of becoming, in which being is never found. These include a rejection of any notion of progress, which supposes that the value of the world is different at different times, and justifies its state at one time by reference to its state at some later time. He summarizes these reflections in three propositions:

> 1) Becoming has *no final state*, does not flow into a 'being'.
> 2) Becoming is *not* an *illusory state*; perhaps the world of *being* is an illusion.
> 3) Becoming has the same value in every moment; the sum of its value always remains the same; *in other words, it has no value at all*, for there is lacking anything against which to measure it, and in relation to which the word

> 'value' would have meaning. *The total value of the world cannot be evaluated*; consequently philosophical pessimism belongs amongst comical things.[42]

These theses have a common theme. They assert that there is nothing that possesses any privileged claim in relation to the world of becoming, whether to finality (by coming *after* becoming) or to reality (by being located *behind* becoming) or to value (by standing *above* becoming). The first case is a state of permanent rest which the world is bound to reach at some future time. We will shortly look at Nietzsche's response in more detail. The second thesis dismisses the idea of an underlying basis for becoming in a 'real' or 'true' world: the Platonist doctrine that, combined with Christianity, has dominated Western thinking. On the contrary, Nietzsche replies, all ideas of such a reality are derived from becoming through falsification and invention. This also points to his third thesis, that there is no standard against which the value of the world can be measured. That comment parallels a Kantian argument about the impossibility of assigning any temporal or spatial magnitude to the world as a whole. Such measures can only be relative, taking some process within the world as their standard in an arbitrary way.

So far, absolute becoming has only been given a first definition, as a becoming that has no beginning, ending or temporary pause that would imply a boundary with being or non-being. More debatable is the assertion of several writers that absolute becoming must always proceed at the same rate. On the face of it, differences in speed would not imply discontinuities. Yet they raise problems of their own about space and time, as seen in Zeno's paradoxes. The other characterization of absolute becoming is a sweeping dismissal of being itself, quite apart from any imagined interface with becoming. Absolute becoming must be the sole reality. It has no explanatory relation to being, as the tradition of Anaximander holds, and a being without any relation to appearance would be what Nietzsche calls 'useless and superfluous'.[43] One of these points provides a way forward for our analysis: the statement that becoming can never come to an end. By looking into the reasons that Nietzsche gives for this thesis, we can raise the question of the inner nature of becoming and try to identify what there is *about* becoming that has this consequence, as well as other implications.

The Impossibility of a State of Rest

For some contemporaries of Nietzsche, notably Eduard von Hartmann and Philipp Mainländer, the idea of an end of becoming has a teleological sense. They believe that the world as a whole has a goal. The aim of the 'world-process' is to renounce the restless and never-satisfied willing that gives rise to the ongoing struggles of life, and hence to all its sufferings. Mainländer, who preserves much of Schopenhauer's metaphysics while lacking his spirit of defiance, calls this 'redemption'. He advocates a voluntary renunciation of life by the human race, not only for itself but on behalf of all living things. However, most of those who took a final state seriously were thinking within the framework of mechanistic materialism and the recently formulated second law of thermodynamics. The question of becoming and being is reopened in the idea of a final state of force, the universal equilibrium of temperature in the universe predicted by physics and much discussed in the popular science of the time.

For Nietzsche, this is a crucial issue for addressing the question of materialism. If the orthodox Newtonian model of physics, extended to include electricity and magnetism as well as thermodynamics, can be shown to lead to a false conclusion, the way is open for a reconsideration of physical theory. He takes the impossibility of a final state to be more certain than any of the premises from which it is deduced. Hence, the mechanistic world view is on trial in any debate over a standstill of force. An 1885 note sets out a line of thought supporting this case:

> If the world had a goal, it must have been reached. If there were for it some unintended final state, this also must have been reached. if it were in any way capable of a pausing and becoming fixed, of 'being', if in the whole course of its becoming it possessed even for a moment this capability of 'being', then all becoming would long since have come to an end, along with all thinking, all 'spirit'. The fact of 'spirit' as a form of becoming proves that the world has no goal, no final state, and is incapable of being.[44]

This is evidently a piece of argumentation, relying on premises from which inferences are made by means of standard logical procedures. The best way to understand it is as a *reductio ad absurdum*, a line of reasoning leading to a conclusion – that the present state of things is not one of change – which we know to be false. Given this outcome, we need to go back to the premises and determine which is responsible for the argument's unsoundness. Nietzsche believes that we will be forced to give up the assumption that a final state is possible.

Setting out the logical form of the argument inevitably involves some reordering of ideas compared with the text quoted above. The first stage relies on two premises: that past time is infinite, and that if past time is infinite, whatever is possible must have occurred already. The conclusion drawn is that whatever is possible must have occurred already.[45] The same starting point is crucial to another argument found in Nietzsche's work, one whose conclusion is that every event that occurs does so not just once, but infinitely many times.[46] So we are really considering both of these lines of thought in looking at the first two premises. Each of them is controversial. The infinitude of time is elsewhere asserted by Nietzsche, and even argued for at length in one text which is familiar to readers, since his editors placed it conspicuously near the end of *The Will to Power*.

> As little familiar as I am with what is philosophised these days amongst Germans: I have discovered, thanks to some fortunate accidents, that in Germany it is now in fashion to think, not of a creation of the world, but at least of a beginning: one resists an 'infinity behind' – But do you understand my abbreviated formula? Mainländer, Hartmann, Dühring etc. all agree on this. The most objectionable expression for the opposite view, that the world is eternal, has been found by Mainländer, an apostle of unconditional purity, like Richard Wagner.[47]

This note addresses an old controversy over the concept of a past eternity, revived by several of his contemporaries, appealing to the quasi-Aristotelian principle that 'an infinity cannot be traversed'. Nietzsche is hostile to attempts to show the impossibility of an infinite past time, because he identifies them as bound up with the religious doctrine of creation. This is certainly true as a historical observation. The argument was first proposed by the Christian theologian John Philoponus.[48] It was then taken up by the Islamic *kalām* school of theology and by Christian thinkers such as St Bonaventure. The nineteenth-century writers such as Eugen Dühring whom Nietzsche had in his sights drew no religious inferences from a beginning of the world, but that silence left them with a puzzling situation: an unexplained transition from original timelessness to continual change.

Nietzsche's dismissive response to the finitist argument is, 'Nothing can prevent me from reckoning backward from this moment and saying "I shall never reach the end"; just as I can reckon forward from the same moment into the infinite.'[49] By itself, the capacity to count backward endlessly is not in question. The issue is whether it establishes the conceptual coherence of a past infinitude. The opposing argument is that the forward direction of the temporal process requires us to count in that direction, and the difficulty is that counting requires a beginning, but in an infinite past there is no such starting point. Here two responses are possible. St Thomas Aquinas accepts this condition, but notes that in every particular instance, the process of counting will start at a finite distance from the present: 'But whatever day from the past we pick on, between that and the present day there is only a finite number of days, which can be traversed.'[50] Nietzsche, in contrast, separates counting from what is being counted, and repudiates any claim that the process of counting must reflect a corresponding aspect of reality in any way other than a one-to-one correspondence of numbers and items. Whether what is being counted is a collection of coexisting things or a succession of events makes no difference. With that in mind, he warns against taking 'the head – this moment – for the tail'.

Here we can at least say that Nietzsche states his reasons, and also that for once his view is consistent with a philosophical consensus, given that belief in finite time has always been a minority position. The case is different with the claim that whatever is possible must occur in an infinite time. As Nietzsche would have known, this thesis was debated among ancient thinkers, for whom it was identical with the fatalist doctrine that whatever happens does so of necessity. Nietzsche clearly accepts the principle, and uses it to rebut the notion that the universe is some kind of living thing: 'If the universe were capable of becoming an organism, it would have done so.'[51] Yet he does not explain why the inference from possibility to actual occurrence is a valid one. This is a puzzle for the interpreter, although several possible solutions suggest themselves.

One is that Nietzsche could be following those ancient thinkers who related 'possible' and 'necessary' to existence in time in a direct way. In their view, the possible *is* just what is actual at some time, while the necessary *is* just what is actual at every time. Each of these propositions is a consequence of the other, given the logical relation between possibility and necessity. These ideas are particularly associated with the philosopher Diodorus Cronus, whose 'Master Argument' was designed to prove, according to Cicero, that 'only what either is true or will be true is a possibility'.[52] Nietzsche would have been aware of Cicero's version of the 'Master Argument' as well as the account given by Epictetus. The argument has exercised commentators, and it is not clear how it is to be understood.[53] The Diodorean position on possibility and occurrence is attractive in its directness. The trouble is that it is not very plausible, for it goes against the common sense assumption that a possibility may never be realized. More consistent with that belief is the definition of Alexander of Aphrodisias: 'It is possible for that to come to be which is not prevented from coming to be by anything, even if it does not come to be.'[54] So there are several levels of uncertainty in this reading of Nietzsche's conception of possibility.

Alternatively, he could be thinking of an argument relying on arithmetical reasoning. Supposing the world to consist of a finite number of elements, it may be argued that the number of combinations they form is also finite, so that a sequence of these could not continue forever but would eventually repeat a previous state. Several notebook passages suggest that Nietzsche did have this in mind.[55] He certainly endorses the argument's finitist premises, although giving only sketchy reasons. The mathematical aspect of the argument is not in question, accepting the 'Archimedean' principle that if a finite quantity (such as a line of a given length) is added to itself repeatedly, it will exceed any given quantity. However, the assumptions made about the world look even less plausible than the conclusion they support. One may hold the world to be finite in extent, and to contain a finite number of parts or elements at any given time, without also holding that the possible combinations or configurations of these are finite in number.[56] The gap between these claims is never addressed by Nietzsche.

Yet another reading of his argument has impressed many English-language writers on Nietzsche since its appearance in a 1960 article by Milič Čapek.[57] Possibility, actuality and infinite time are brought together here by invoking the concept of probability and the principle that any event with a finite probability will occur in an infinite number of trials. That is, its non-occurrence will become, if not impossible, at least infinitely improbable over a long time. Čapek cites Hans Reichenbach's illustration: 'When we throw a die often enough, a run of a thousand throws showing face 6 on top must eventually occur, because the probability of such a run, though very low, is larger than zero.'[58] The example can be taken further: if we go on throwing the die after this run has occurred, it must eventually occur yet again, and do so infinitely many times. Now, it is assumed here that the run specified has a finite probability, on the further assumption (based on experience) that a single throw of 6 has a finite probability. Whether that can be said of the present state of the world as a whole – the case used in establishing an eternal return of the same – is another matter altogether. How do we know that this state of affairs has a finite probability of occurring? The fact that it has occurred proves that it is possible, but says nothing about its probability. And what sort of property is probability, anyway? Certainly not an observable one. The theory of probability provides no simple solution and in fact opens up further disputed issues. The most common approach derives probability from empirical information about the frequencies of different outcomes. So, are we to keep track of the occurrences of a particular total state and compare their frequency with that of other states? This will not be easy, if our counting is occurring within the actual sequence. Yet we can spare ourselves the trouble. For if every total state is determined by the previous state, and in turn determines the next state, then their frequencies must be exactly the same, and so must their probabilities, if this is the only basis for assessing them.

I think that we can avoid all this. The argument from probability is not required, since it includes everything that the previous argument assumes. If every possible state of affairs has a finite probability, as assumed here, then there can be only a finite number of possible states of affairs – and so the Archimedean argument can proceed as before, without the distraction of issues concerning probability.

The next stage of Nietzsche's argument introduces the premise that a state of rest is possible. Nietzsche does not bother to mention this, presumably because it has already been put forward by the current physical theory that is the background of the discussion, and because it will come in for closer attention as the target of the *reductio ad absurdum* argument. The final stage of the argument depends on a new premise: that if an overall state of rest were to occur, there would be no further change afterwards. The rationale is the principle of sufficient reason, but the formulation echoes arguments used by Galileo against the Aristotelian physical theory that requires a state of rest between distinct motions. One of Galileo's illustrations concerns moving objects that collide with each other, so that one changes its direction of motion. He writes, 'But how can they say that at that point of impact an interval of rest occurs? For if they once remained at rest, they would thereafter always be at rest, since they would not have reason for moving.'[59] Nietzsche's argument against the possibility of any standstill of force is the same. However brief in duration we suppose this pause to be, it would necessarily become the permanent state of things, because there would be no reason for further change to occur.

Putting these steps together, we come to the conclusion that the present moment is not a state of change. And that, Nietzsche says, we know to be false. This is where he puts the proposition that the present moment is one of becoming to real use: 'The fact of mind (*Geist*) *as a becoming* demonstrates that the world has no goal, no end state and is incapable of being.'[60] Of course, this 'fact' (a rare claim for Nietzsche) does not justify such a conclusion all by itself, but when further premises are added it does support a demonstrative argument for that thesis.

Why Is a State of Rest Impossible?

Proving the impossibility of a state of rest is one thing, but explaining it is another. Nietzsche considers three possible answers to this question. One is that the course of becoming is guided away from a state of rest by some overseeing power that makes a timely intervention whenever this state is approached too closely. The second answer is that the nature or 'essence' of force itself implies continual change. The third is that change proceeds in a recurring cycle from which any state of rest is excluded. Each of these responses needs careful consideration, since they contain hidden complexities.

It is true that Nietzsche dismisses the first idea out of hand and with sharp hostility. The hypothesis of a supreme power raised over the world of becoming and controlling its progress, either to attain some goal or to avoid a particular outcome, as in this case, is inconsistent with his rejection of all forms of otherworldliness, in the interest of values promoting life. From a theoretical standpoint, a belief in providence would undermine attempts to construct a scientific model of the world by preventing the principle of naturalistic explanation from being applied comprehensively, even as a regulative hypothesis. At best one can say that invoking a *deus ex machina* in order to avoid a final state is less objectionable in Nietzsche's eyes than doing the same thing to avoid an eternal recurrence. Of the latter notion he is especially scornful. The world, he argues, cannot give rise to an endless supply of new states:

> If it were *eternally becoming new*, it would thereby be posited as something in itself *miraculous* and godlike, free and self-creating. Eternally becoming new presupposes: that force itself voluntarily avoids and has not only the intention but also the means to protect itself from repetition, returning to an old form, thus controlling every motion in every moment for this escape – or else the *incapacity* to return to the same state: that would mean that the sum of force is no something fixed, and similarly the properties of force. Force as something un-fixed, something undulatory, is quite unthinkable for us. If we do

> not want to phantasise into what is unthinkable or fall back into the old concept of a creator (increase from nothing, decrease to nothing, absolute arbitrariness and freedom in development and in properties) – [61]

There are a number of aspects to this idea of *avoiding* a repetition: notions of intention, of voluntariness, of consciousness and of an infinite capacity to generate new forms. Most of these (although not the last one) also apply to any avoidance of a state of rest. Now, that idea may seem easier to justify than the claim that force is incapable of repeating a previous state. After all, a state that has existed is plainly possible, or at least has been possible. And we know there are many such states, whereas there might be just one that would constitute an equilibrium of force. Avoiding a unique situation is an easier task than avoiding a great number of diverse cases.

So, what is required to avoid either a state of rest or the recurrence of a previous state? Sometimes Nietzsche suggests that this would involve an all-knowing and all-powerful God, ruling over the world. At other times he understands it to mean attributing consciousness and intentionality to each of the basic elements that together constitute the world. In the same notebook he writes, 'Monads which know how to avoid certain mechanically possible outcomes such as equilibrium! Phantasising!'[62] Either way, the foresight attributed to the constituent elements amounts to the sort of anthropomorphism that Nietzsche campaigns against in his natural philosophy. We need to 'dehumanize' nature to achieve a consistent naturalism, he says, even if that means giving up the familiar notion of causal efficacy as modelled on human agency, itself the target of a destructive critique.

Having refused to accept any teleological answer to the question why a final state is impossible, Nietzsche considers two further responses in an 1881 notebook entry: 'Complete equilibrium must either be an impossibility in itself, or the alterations of force enter into a circular course before the occurrence of that equilibrium, which is possible in itself.'[63] The first idea is that the nature of force makes a state of rest impossible. Elsewhere he spells out this option:

> 'Timeless' to be repudiated. In every particular moment of force the absolute conditionality of a new division of all its forces is given: it cannot stand still. 'Alteration' belongs to its essence, and thus temporality as well: but in that way the necessity of alteration is just conceptually posited once again.[64]

The second hypothesis is that the flow of becoming happens to enter into a circular course before a state of rest is reached, so that this state is *prevented* from occurring from that point onward. Both ideas are consistent with Nietzsche's standpoint, but the first answer, identifying the nature of force with continual change, is his preferred one. As he puts it, '"force" and "rest", "remaining the same", contradict one another'.[65] We should note here the use of 'contradict', which will appear again in his model of time involving lanes which 'contradict' each other. Nietzsche is not thinking primarily of verbal formulations. If he were, the proposition that 'force' and 'rest' are inconsistent would be hard to justify, since neither term lends itself to further analysis. Rather, his explanation would be through the concept of conflict, not only essential to force but entailing continual change: a theme to be discussed in the next chapter.

The second option is of interest, and one wonders why Nietzsche did not take it more seriously. A doctrine of eternal recurrence has important consequences for our understanding of what is possible and impossible. Future events, on this hypothesis, can only be those events that have occurred in past cycles of becoming. One may imagine events other than these ones, which are ruled out by nothing other than the fact that they do not belong to that particular set of events. There is a good deal of scope for contingency in this restriction on what is possible. Of course, some events can be identified as impossible from description alone, since they are inconsistent with laws of nature which impose their own necessity on the course of becoming (to use a language that Nietzsche would prefer to avoid if he could). Apart from that consideration, if we do not know whether a given event has occurred before, then we cannot say whether it is possible or impossible.

This is a startling enough conception: that a state of rest is quite possible, regarded by itself, but by pure chance has been ruled out once and for all, because the universe happens to have fallen into a circular course. It seems natural then to say that the state of rest is now impossible, since it is certain that it will never occur, although it was possible up to the point where the circle of events was closed. To use a twentieth-century analogy, the world is like a gramophone record that has become stuck in the groove while playing, so that its ending – otherwise necessary and predictable, being impressed on the disc – will never be reached. This illustration is inaccurate in one respect, because it involves a process preceding the circular course. In fact, it presupposes a beginning, for otherwise the argument for the necessary realization of every possibility in an infinite time would apply. If there is no beginning, then the circular course must always have existed. And if recurrence has been going on for a past eternity, then there was never a time at which the final state, or for that matter any other one not belonging to the cycle, *became* impossible. These states were always impossible, despite being possible 'in themselves'. They were prevented from occurring by the cycle's designation of certain states as the only ones that occur. In that case, we cannot define this impossibility in terms of 'becoming impossible', although in the absence of an eternal recurrence that would be its typical explanation.

Why did Nietzsche not accept this conceptually economical answer to his question about a final state? Possibly the contingent status of this version of eternal recurrence was unappealing. More probably, his bias in favour of a doctrine of becoming, and his wish to make that principle the springboard for a comprehensive conception of reality, made an alternative seem already available on these grounds. A conception of force going beyond the mechanistic world-interpretation (which is all that the 'accidental' explanation of constant change needs) promised a powerful model for use not just in physical explanation but in accounting for human phenomena as well. Hence, Nietzsche's references to an 'essence' of force, characterized by continual change, despite the quantitative constancy of force as a whole. The next chapter will address this 'essence' of force, in the hope of discovering how it not only entails absolute becoming but also provides the basis for its interpretation as time.

Chapter 2

From Becoming to Time

I said earlier that for Nietzsche, becoming is a fact, whereas time is an interpretation. The task now is to explore this process of interpretation and, in particular, to identify the aspects of becoming that make it possible. That is what this chapter will attempt to do. There are some puzzles about this undertaking. It requires us to make sense of a becoming that is not – or not yet – identified with time. What can we say about this that will not presuppose the sort of interpretation whose possibility we are trying to account for? Similarly, how can we describe the process of interpretation without imagining a time in which it takes place? If there is to be an interpretation of becoming, one could say, there must also be a becoming of interpretation. The origin of time is not an issue for Kant's philosophy, according to which time is given all at once in a single intuition, and not constructed in the understanding.[1] But what about time as an interpretation? That is a process that happens and, like anything that happens, it seems to require time of some kind. As Nietzsche puts it in one notebook entry, 'Our derivation of the time feeling etc. always still *presupposes* time as absolute.'[2] In another note he observes: 'To the actual course of things must also correspond an *actual* time, quite apart from the *feelings* of longer and shorter intervals that we knowing beings have.'[3] These working notes may present points of argument that have occurred to Nietzsche rather than his own judgements. Still, it is hard not to suppose that there must be a real time, even if it is unknown to us, underlying the processes that go on in our minds. So, is there a paradox in any attempt to explain time concepts as the outcomes of a process of interpretation?

Several Nietzsche scholars have speculated on the concept of a time that must stand outside experience, since it is where the origin of experience is found. John Richardson proposes a contrast between the time within perspectives – that is, the time that they are 'of' or 'about' – and their own temporal structure or becoming.[4] His discussion starts from a related issue: if we attribute the doctrine of becoming to Nietzsche, should this be understood in a realist sense? Nietzsche often expresses scepticism about all claims to givenness, and this would extend to the most general features of reality. Further, it is not easy to grasp becoming as a feature of reality 'in itself'. Richardson argues for a position that avoids this dilemma by locating becoming in the nature of perspectives, although not in their content (which is, in fact, more likely to amount to a negation of becoming).[5] He places this conception within a naturalistic setting: only living things can have a perspectival temporality of past, present and future, because that arises out of drives that are temporal in nature. On the one hand, drives are directed towards particular future goals, often described by Nietzsche as 'values', and their striving towards these is itself valued as an overcoming and gain in power. At the same time, drives are what they are because of a past process of natural selection. Their aims express the process that has brought them about, the object of what Nietzsche calls 'genealogy'. In contrast, the past is presented within the perspective of the organism either as of no importance (and hence, simply forgotten) or as a mere means to achieving its given aims.[6]

Richardson is writing about the origin of the perspectival structure of consciousness. But there is also a question about any particular experience and its origin, that is, about how whatever is given comes to appear within that perspective. This must be a process of some kind, so can we speak of time there? Werner Stegmaier draws a contrast between what he calls the *Zeit der Vorstellung* and the time that we experience, the *Vorstellung der Zeit.* He argues that this first, 'other' time must necessarily be ungraspable by the human mind, because our mode of representation imposes a 'pre-given mechanistic time-schema of simultaneity and succession' which is a falsification of absolute becoming.[7] For the same reason, we cannot even show that 'there is' such a thing, or treat it as a single topic.[8] Despite this negative conclusion, Stegmaier goes on to argue that the thought of eternal return directs us towards the 'other' time located outside representation. Its conceptual intractability undermines the claims of representation: the will to power's demand for a continual transformation through interpretation cannot be reconciled with an affirmation that wants nothing to be other than it is. Hence, Stegmaier concludes, we are compelled to confront the ultimate failure of representation and, if only by indirection, to acknowledge the 'other time' from which representation arises in the first place.

Both writers draw a sharp distinction between experience and what underlies it. If we look at Nietzsche's expressed views, the situation is not so clear. It is true that he often insists on the impossibility of getting away from our conceptual framework, but he also offers clues about the actual process of interpretation, one of whose outcomes is our conception of time. What is happening when the mind constructs a temporal model on the basis of the becoming that it encounters? What is there about becoming that makes it possible for us to identify periods of time, or to distinguish between past, present and future? One way of approaching this process is by looking at the instances where it does *not* occur in the usual way. These anomalous cases allow us to get behind the scenes, so to speak, and identify what is taken for granted in everyday experience. For Stegmaier, the experience of eternal return is the paradigm case, owing to what he takes as its paradoxical character. Nietzsche too speaks of it as a radical challenge to thought, but he has other experiences in mind as well. It will be instructive to consider some that he discusses and what he thinks can be learned from them.

Anomalies of Time

Can we find clues within experience to what underlies experience? Nietzsche takes an interest in some mental states that he thinks throw light on the construction of time. Often these have to do with the *tempo* of becoming, as when he observes that hashish smoking has the effect of making time slow down, so that 'much more than usual is seen in the same time'.[9] Many of his examples are concerned with sleep and dreaming, or with the borderland between sleep and waking that is commonly called 'insomnia', something with which Nietzsche was familiar throughout his adult life. The ridicule directed against the complacent eulogists of sleep by Zarathustra in 'On the Teachers of Virtue' is mild: it is not his intention to query the worth of sleep, only its use as a criterion for assessing behaviour. Despite the sedative drugs that he took without medical supervision and which his sister Elisabeth blamed for his eventual mental breakdown, Nietzsche's frequent bouts of sleeplessness are in evidence in many passages of his writing. In one note of 1881 he writes:

> There is a part of the night about which I say, 'Here *time* ceases!' After all night wakefulness, especially after night journeys and walks, one has a strange feeling in relation to this period of time: it was always much too short or much too long, our sense of time feels an anomaly. It may be, too, that we have to pay for it upon waking, given that we usually spend that time in the time-chaos of dreaming! Enough, at night from 1 to 3 o'clock we no longer have the clock in our head.[10]

He cites several expressions used by classical authors for this part of night, such as *nocte intempestiva* (more correctly, *nocte intempesta*), meaning literally 'untimely night' but glossed by Nietzsche as 'there in the night, where there is no time'. The Latin phrase refers to the time after people retire to sleep, often carrying an implication that anyone who is still out and about then is probably up to no good.[11]

What interests Nietzsche are the 'anomalous' feelings linked with states of mind that lie between waking and sleeping: the experiences of 'overawake souls'.[12] Sometimes he associates this disoriented condition with drunkenness.[13] The link is evident in 'The Drunken Song', the conclusion of the 'Ass Festival' episode in *Thus Spoke Zarathustra.* In the first edition this chapter was called 'The Sleepwalker's Song' (*Das Nachtwandler-Lied*), underlining the theme of night and its associations. But why 'The Drunken Song'? Despite the festive setting, including wine, the participants are not shown as drinking to excess. It is the water-drinking Zarathustra who is described as standing 'like a drunkard'.[14] Clearly this expression is a figure of speech. More importantly, it points to an affinity with themes of *The Birth of Tragedy.* There Nietzsche had identified drunkenness as an analogy to the Dionysian impulse, in contrast with dreaming, taken as corresponding to the Apollonian drive.[15] If this contrast is applicable, can the interpretation of becoming as time be seen as an instance of the Apollonian drive towards form? Consistent with this is its invention of regular structures within which events or states of affairs can be organized and treated in relation to one another. That schema is disrupted by anomalous experiences of time or of 'timelessness'. The silence of the 'clock in our head', and the absence of the 'wide awake day-wisdom'[16] that takes everything to be countable or measurable, allows the 'time-chaos' that underlies structures of time to emerge in awareness.

In several episodes Zarathustra experiences a suspension of any sense of time during these altered states of consciousness. In 'The Soothsayer' he describes a dream where 'time passed and crawled, if there was still any time – how should I know?' In 'The Stillest Hour' he reports a dreamlike state in which time seems to stand still: 'The hand moved, the clock of my life drew a breath . . .' As described, these brief pauses are everyday occurrences. When a clock's hand moves, it commonly stays where it is before moving again. Similarly, drawing breath is a familiar action in speech, as well as in singing or playing a wind instrument. If a pause occurs, it is so inconspicuous that it usually goes unnoticed. Yet it is of interest to Nietzsche, who notes in an essay on rhythm and meter that the poetic caesura has its origin in 'the brief resting moment of drawing breath'.[17] These are apt metaphors for what he is trying to indicate: a suspension of time which is not, however, the *negation* of change that gives rise to a belief in enduring things and the 'I' that corresponds to them, in turn supplying the content of structured time. Rather, a pause of this kind is a 'time out' – that is, outside time, which is not to say outside becoming.

Nietzsche gives this concept an important epistemological function. It is something done by the mind as change continues in the rest of the world, and the consequences are far-reaching:

> Our consciousness limps behind and observes little at one time while it pauses for something else. This incompleteness is surely the reason that we believe in things and assume something enduring within becoming: it is also why we believe in an 'I'. If knowledge were to run as quickly and as continuously as the course of development, there would be no thought of an 'I'.[18]

Why does consciousness behave in this way? As described, it looks like a weakness. The human mind is unable to keep up with the pace of reality. It lags behind because it is trying to deal with some impression, to interpret it by finding associations that can be accepted as an explanation. Biologically speaking, this is a favourable adaptation that serves the need to assimilate impressions by suspending judgement in the meantime. An organism that reacts immediately to every stimulus is less likely to cope successfully with its environment than one that selects and evaluates *before* an impression enters conscious awareness. Nietzsche's favoured analogy is with nutrition: every living thing needs to judge what is good or bad to eat – or to know. This suspension of reception needed for the processing of sensory input into the format required by our mind is characterized by Joan Stambaugh as a timelessness which is also the origin of organized succession in time.

> The time that precedes all experience and motivates it, makes it possible, is what forms timelessness in contrast to an inner state. Strictly speaking, timelessness does not stand in contrast to the inner state, but first constitutes it. This 'timelessness' is not so called because it is static but because it is the initial moment of the occurrence of time, which first makes possible an encounter with something. It is 'timeless' in the sense that it is not yet the elapsing succession of moments at rest separated from each other. Succession first arises from timelessness. Timelessness is 'true', i.e. primordial time which first makes possible appearance or error.[19]

One must note, however, that this is not a pause in the perceived course of the world. Hence, it is consistent with Nietzsche's insistence that the present moment is one of becoming. The time of the pause (if we can call it time, although it has no definite length) is not the time of the world. What is happening is an interaction between the outer stimulus and the inner perceptual apparatus. The forces that make each what it is come together and enter into an interaction. That implies time since, as Nietzsche puts it, 'Forces take definite times to become definite qualities.'[20] He is puzzled by the idea. Does it imply an absolute time scale – and in turn, a given unit of time? More plausibly, this is the unconscious process by means of which an impression arrives in consciousness. Every quality is originally the symbol of some such transaction, a simplified signal to the sentient creature, although over time we form habits of identifying the 'same' quality in similar cases – again, a favourable adaptation for coping with a complex environment where quick decisions are needed.[21]

Closely associated with this idea of a pause or delay in our awareness of the world is another, more paradoxical concept. Nietzsche speaks of an 'inversion of time', meaning an opposition between the order of successive events in consciousness and that of the underlying process. What makes this idea plausible for him is an example which seems to prove that we already do something of the kind: the incorporation of an external stimulus within a dream experience.[22] Hearing a cannon shot while asleep, Nietzsche says, we experience it within a dream, but interpolate a fictitious process leading up to and producing it (or whatever corresponds to it).'The cannon shot appears in a *causal* mode, in an apparent reversal of time. What is really later, the motivation, is experienced first, often with a hundred details that pass like lightning, and the shot *follows* . . . What has happened? The ideas that a certain state *produced* were misunderstood as its cause.'[23]

This phenomenon supports a broad epistemological thesis about 'imaginary causes'. Nietzsche boldly suggests that the same confabulation is very common in waking life. The supposed cause of a sensation in the external world is usually our own work, initiated by that impression before it reaches awareness. The actual datum is projected into the external world, where it can then appear as a cause that works on us. 'It takes time, before it is ready; but this time is so small.'[24] Still, confabulation must itself be a process, if not a conscious one: it is a 'pause' only in contrast with the further course of external events. Nietzsche adds that the object is meanwhile held before us as the same thing: presumably this is the real origin of our belief in persisting things. Further, the counterpart of this enduring object is a perceiving subject, and this is the 'I' that we take to be equally enduring.

What do we learn from these anomalous cases? Nietzsche thinks they show that succession is not given, as empiricist thinkers tend to assume, but is constructed in the mind. The outcome is therefore not something that should be taken at face value. Rather, it must be subjected to a critique that places in question not only the basic framework of time, but also the concept of causality. In looking more closely into Nietzsche's treatment of these themes, we will also be encountering his alternative, grounded in an acknowledgement of *conflict* as the basic character of becoming.

Succession, Causality and Conflict

In approaching the interpretation of becoming as time, Nietzsche addresses an account that he thinks is the common and accepted one, centring on the concepts of succession and causality. In an 1881 note he writes, 'Only succession produces the idea of time. Assuming we did not sense causes and effects but a continuum, we would not believe in time.'[25] Later in the same note he returns to this hypothetical situation. What would our idea of reality be if we were to perceive the course of becoming not as a sequence of distinct events or states of affairs, but as an uninterrupted flow?

> A continuum of force is without succession and without coexistence (this too presupposes the human intellect and gaps between things). Without succession and coexistence there is no becoming for us, no multiplicity – we could only assert that continuum to be one, at rest, changeless, not a becoming, without space and time. But that is only the human antithesis.

This can be seen as a radical critique of our ordinary concepts of space and time. Nietzsche is saying that they depend on discontinuities or 'gaps between things' which are not present in becoming, but are introduced in the first place by the human mind. That raises a new question: how does the mind order these events or states in a single succession, as occurring earlier and later in time? The anomalous cases provide a clue: the earlier events are supposed to be the *causes* of the later ones. They provide explanations that give the mind a claim to achieve a grasp of becoming. But the validity of that claim is open to question.

Nietzsche mounts a radical attack on the categories of cause and effect, and proposes a different way of locating the basis of succession in becoming. We can see both this attack and his alternative in one passage that not only identifies becoming with conflict but gives this aspect an explanatory role in the emergence of time.

> Every conflict – every occurrence is a conflict – *takes time* (*Aller Kampf – alles Geschehen ist ein Kampf – braucht Dauer*). What we call 'cause' and 'effect' leaves out the conflict and consequently does not correspond to the occurrence. It is consistent to deny time in cause and effect.[26]

Here the idea of conflict is linked with time in a way that places it in opposition to cause and effect. The concept of causality, despite its success in explaining and predicting events, leaves time an insoluble mystery. A causal account cannot say why a certain period of time should elapse between a cause and its effect, or any time at all. As Nietzsche points out, the immediate occurrence of the effect would be consistent with the explication of a cause as a sufficient condition. What could time by itself add to a cause that is already present? Worse still, if causal action does take time, how can the effect be guaranteed, given that some other factor may intervene during the time period and prevent it from occurring? Causality raises more questions than it answers where time is concerned.

The alternative is found in the passage's opening argument: it is because all conflict takes time that everything does not happen at once. Part of Nietzsche's debt to ancient Greek thought is the principle that becoming is always a contest of opposing forces. He can claim the authority of Anaximander and Heraclitus, among others, for this conception. The struggle between heat and cold, light and dark, wet and dry, bitter and sweet goes on forever. If it were possible for one side in these conflicts to achieve a permanent victory over the other, that would long since have occurred. Nietzsche links the assertion of Heraclitus that all things are produced by war (*polemos*) with the Greek view that competition – the good goddess of strife (*Eris*) in Hesiod's *Works and Days* – is the source of all striving for achievement in human society.[27] It is important to realize that in his view, conflict is not something destructive, but the chaotic and accidental source of creativity. We need to see now how these concepts are related and brought together.

Conflict *within* the individual person is especially critical for Nietzsche. He writes, 'This is the deepest conception of suffering: the formative forces collide with one another (*stossen sich*).'[28] That is painful because the collision of forces causes frustration to the creative will. And yet, Nietzsche says, achievement requires just this pain in order to be an overcoming of resistance. 'Displeasure, as an obstacle to its will to power, is therefore a normal fact, the normal ingredient of every organic event; man does not avoid it, rather, he is in continual need of it: every victory, every feeling of pleasure, every event, presupposes a resistance overcome.'[29] One text is especially instructive in showing how Nietzsche associates becoming with inner conflict. It occurs in an 1888 note headed 'Morality of decadence', discussing human attitudes towards the senses and passions, and the common tendency to fear their strength. Nietzsche describes three possible situations in personality. One is the domination of a ruling passion, which he suggests is a state of complete health, since it maintains a stable organization. Another is an unhealthy struggle between passions, leading either to general ruin or, on the best outcome, to the emergence of one drive as a dominating influence. The third possibility is this:

> Juxtaposition (*Nebeneinander*) without being antagonism or collaboration: often periodic, and then, as soon as an order has been established, also healthy . . . The most interesting people, the chameleons, belong here; they are not in contradiction with themselves, they are happy and secure, but they have no development (*Entwicklung*) – their states lie juxtaposed, even if they are separated sevenfold. They change (*wechseln*), they do not *become.*[30]

Here 'becoming' is identified with development and growth, a process occurring through both conflict and co-operation, subordination and co-ordination. 'Change' is a weaker concept, suggesting alteration but not any particular process or direction. Why does Nietzsche refer to people whose drives coexist without conflict as 'chameleons'? In *Daybreak* he argues that social morality is already found in the animal world, and gives an example: many creatures adapt their colouring to match the colour of their surroundings, in order to go unseen by their natural enemies.[31] So too in human society, people conform to the attitudes, beliefs and customs of those about them, in order not to draw attention to themselves. With this in mind, Nietzsche's attitude to the 'interesting' people that he calls chameleons should not be mistaken for admiration. The changes that he attributes to them, however dramatic they may seem, occur only on the surface, and leave what is important untouched.

So, why does he say that they do not *become*? That observation could be read as, 'they do not *become what they are*'. The phrase is sometimes used as an injunction or maxim, to 'become what you are'.[32] That the word 'becoming' acts as an abbreviation for this stronger idea is confirmed by a reference in *Ecce Homo* to 'my innermost history, my *becoming*'.[33] It looks now as if 'becoming' means not just alteration, but the kind of metamorphosis of the spirit described in the first chapter of *Zarathustra*, radical enough to be an equivalent of coming to be and passing away. How does that development take place? Nietzsche's answer appears in this passage: by interaction between the opposing forces within the individual person: 'the *collision of the creative impulses*'.[34] If these merely coexist, as in the people he has described, there is a benign situation, but no genuine becoming, because no self-overcoming. 'Becoming what you are' has a paradoxical character because it combines the language of open-ended development with that of essential forms or types. It hints at an overcoming of the contradiction of being and becoming, but it would be more accurate to say that Nietzsche has in mind an image of being within the world of becoming.

Accidentality and Chaos

This conflict model of becoming has many implications. Take, for instance, the notion of necessity. Again, Nietzsche suggests a thought experiment: what would our view of the world be if we did *not* impose 'gaps' on becoming and then try to reconnect the separated fragments by means of causality? In *The Gay Science* he writes, 'An intellect that could see cause and effect as a continuum and not, as we do, in terms of an arbitrary division and dismemberment, that could see the flow of becoming, – would repudiate the concept of cause and effect and deny all conditionality.'[35] An absence of causality implies the elimination of all claims for responsibility and accountability. Nietzsche calls this the 'innocence of becoming'. He regards causality as an essentially moralistic concept, motivated by the desire to assign responsibility for suffering to some agent – preferably one who can then be made to suffer.

> Wherever responsibilities are sought, it is usually the instinct of wanting to *judge and punish* which is at work. Becoming has been deprived of its innocence when any being-such-and-such is traced back to will, to purposes, to acts of responsibility: the doctrine of the will has been invented essentially for the purpose of punishment, that is, because one wanted to *impute guilt.*[36]

Nietzsche wants to restore the innocence of becoming in his doctrine of complete non-responsibility. But it is not simply a question of the necessity which, after all, is affirmed by supporters of the causal determinism that he opposes. As he points out in *The Wanderer and His Shadow*, the principle of causality universalizes responsibility, rather than eliminating it.[37]

The innocence of becoming is identified by Nietzsche with 'the wholly pure accidentality of occurrence (*Geschehen*)'[38] and 'the enormously *accidental* character of all combinations'.[39] Accidentality is the true character of everything that happens, despite the systematic denial built into our conceptualizations. But just what is accidentality? An accident is a certain kind of event, or rather, a combination of processes that are independent but intersect in a way that produces significant consequences. For Nietzsche, this is a corollary of conflict. When forces come into conflict with one another, what results is an 'accidental' or 'chance' occurrence, as well as a site of creativity: 'I recognised the *active force*, the creative within the accidental – accident (*Zufall*) is itself only the *collision of the creative impulses.*'[40] Another word for this is 'coincidence'. Aristotle recognizes *coincidence* (*to kata sumbebekos*, literally 'things that have come together') as a kind of event with specific features. A coincidence occurs when things that are not usually found together encounter each other with an outcome that can be regarded as an event in its own right. If two people meet by chance, for example, we can say why each of them was in that place at that particular time, but this is not to say why *their meeting* occurred. That event may have its own consequences, but as Aristotle explains, when we trace the causes of those later events, we get to an event which is the limit of explanation, that is, one that cannot itself be attributed to a particular cause.[41] This is the accidental event.

The ability and willingness of the human mind to grasp the accidental are both very limited. Our natural tendency is to interpret events in a way that provides an explanation, even a contrived one. An example is the religious doctrine of providence that eliminates the notion of accidentality altogether from the world of experience.[42] Even within a naturalistic world view, there is resistance to any admission of chance. Sometimes 'chance' is explained as a dummy word that serves to signal our ignorance of something's cause but is mistakenly taken to mean that it has no cause. Boscovich expresses this view when he writes, '"Chance" is an empty phrase without a thing to correspond to it.'[43] On the other hand, the word 'chance' is nowadays associated with the idea of probability. Where a situation is too complex for us to identify the many influences at work (in throwing dice, for example) we give up any attempt at causal explanation and speak of 'chance'. We then look for a calculation of probability as a substitute that will provide some satisfaction to our understanding and even practical guidance. Nietzsche does not think like this. 'What are "statistics" in these things!' he writes, referring to the possibility of a cosmic catastrophe.[44] Hence, although his talk of 'the great dice game of [the world's] existence'[45] encourages the assumption that probability enters into his arguments, this is a misreading. As far as he is concerned, a throw of the dice is neither probable nor improbable, although it may be fortunate or unfortunate.[46]

For these reasons, it is better when reading Nietzsche to translate the German word *Zufall* into English as 'accident' or 'accidentality' rather than 'chance'. This is confirmed by the relevance of the Aristotelian conception of accident as coincidence. Nietzsche's scepticism about the predictability of nature and his rejection of 'laws of nature' are both included in this interpretation. An accidental intersection of separate processes implies necessity, but without any explanation of the resulting event. Since each of the several processes occurs of necessity, we must say that their intersection is also necessary, but that is consistent with the unpredictability explained above. In a note from 1880, Nietzsche argues that the ancient concept of *ananke* refers to 'the realm where everything proceeds arbitrarily (accidentally), where it is not the case that from every cause its effect must follow'.[47] This is closer to his own understanding of necessity than a modern identification with universal causal determination.

Nietzsche's alternative to a causal model is indicated by the words 'conflict' and 'accident', but this needs to be linked with another idea which at first sight looks different: that processes combine with one another by being 'entangled'[48] or 'knotted' together.[49] Nietzsche uses several metaphorical expressions for what is clearly a single idea. The earliest version occurs in his lectures: 'Heraclitus perceived that the contrary predicates draw each other after themselves (*sich nach sich ziehen*), something like what Plato says about the pleasant and unpleasant in the *Phaedo*: they are intertwined as in a knot.'[50] In *Thus Spoke Zarathustra* the image appears several times in relation to the eternal recurrence. Zarathustra predicts his return to an identical life by saying: 'But the knot of causes in which I am entangled recurs and will create me again.'[51] What are 'causes' doing here, if the image of entanglement is, as I have been arguing, an *alternative* to talk of cause and effect? I suggest that this is, in fact, a reference to the way that forces 'become particular qualities' as they interact with one another.[52] The 'I' of sensory experience is an outcome of these processes.

In this way, the metaphor of entanglement is an extension of the image of an encounter or collision between forces. It is the outcome of their accidentally coming together, the new situation arising from the fortunate or unfortunate throw of the dice. In the Epicurean world view, atomic particles move within empty space and collide with one another, forming temporary combinations while retaining their separate existences. Nietzschean processes are not like this. They interact by becoming 'entangled' and turning into what he calls 'complexes of occurrence' (*Complexe des Geschehen*) which are more or less lasting in time, like the elements of becoming out of which they are formed.[53] These metaphors hint at a comprehensive alternative to atomistic materialism. Nietzsche never addresses that at length, and it remains an indication rather than a worked-out theory. Still, it plays a significant role in *Thus Spoke Zarathustra*, as we shall see in Chapter 5.

These themes, along with a hint of the dice game, appear in Nietzsche's description of mental life: '*Chaos* is still continually at work in our own mind: concepts, images, feelings are accidentally brought together, thrown about (*gewürfelt*) by one another.'[54] Out of this continually changing mixture the mind selects some combinations for special attention: the fortunate throws that suggest regular patterns, and make possible not just its creative achievements but also its everyday operations. Nietzsche wants to put the notion of chaos to a much broader use. He writes, 'The total character of the world, however, is in all eternity chaos – in the sense not of a lack of necessity but of a lack of order, arrangement, form, beauty, wisdom, and whatever other names there are for our aesthetic anthropomorphisms.'[55] In a discussion of this Nietzschean concept, Babette E. Babich uncovers a more fundamental meaning by arguing that what is crucial is not the absence of order, but a primordial Dionysian abundance of creative energy. She emphasizes its application not only to nature but also to humanity: Nietzsche's idea of a 'dehumanization' of nature leads in turn to a re-naturalization of human life, recreated as both artistic activity and work of art. 'Hence,' she concludes, 'for the sake of cosmic creativity, for cultural creativity, what is needed is an affirmation of chaos as change, alteration, impermanence, flux. This restored emphasis on unfathomable excess also restores the innocence of becoming.'[56]

Variations in Becoming

The perpetual conflict inherent in becoming was understood by the ancient thinkers as between pairs of opposed qualities: hot and cold, wet and dry, dense and rarefied, and so on. Nietzsche's view is that these dichotomies are not given but created by us.

> If we give up the effective *subject*, then we also give up also the *object* that is effected. Duration, self-identity, being are inherent neither in what is called subject nor in what is called object: they are complexes of occurrence apparently durable in comparison with other complexes – thus e.g. through a variation in the tempo of occurrence (rest-motion, tight-loose: all opposites that do not exist in themselves and with which are actually expressed only *variations in degree* that for a certain optical measure are picked out as opposites).[57]

What are these variations within the course of becoming? The first mentioned are differences in tempo: one can see how the appearance of rest may arise from the application of a relatively short measure of time, as in the thought experiments of the biologist Karl Ernst von Baer, to be discussed in Chapter 4. The other phrase, 'tight-loose' (*fest-locker*), is probably about the way that concurrent processes are 'intertwined' with one another. If these entanglements are less 'tight' in some cases, the processes may be taken as separate – again, a mistaken interpretation due to the limitations of our perception. But what feature of becoming underlies and makes possible these perspectival falsifications? The word 'degree' (*Grad*) is a crucial clue. It indicates a magnitude which is not the extensive magnitude of space, time or mass. The general expression for this is *intensity*.

The concept of intensity is significant in epistemology. Empirical knowledge is grounded in sensation, the given element in experience. Sensations have intensive magnitudes, as do the qualities of the objects that produce them, although the relation between psychical and physical intensities is not a simple proportionality. Sentience arises from interactions with the environment, driven by tension and conflict between forces. It is doubtful whether we ever sense a single force, or forces which balance one another. If the pressure acting on our bodies is the same as the pressure within, we do not experience it as pressure, however great its magnitude. Similarly, we sense heat and cold only with a difference in temperature between our own body and the object with which it is in contact. These differences have consequences: pressure brings about movement, while heat passes from one body into another – which is what produces the sensation.

Since conflict between forces is due to differences between intensities, it is not surprising that what matters for quality is not intensity as such but the differences between intensities. The greater the difference, the more intense the sensation. A sensation of cold can be just as intense as a sensation of heat, although it corresponds to a relatively low intensive magnitude in the objective quality. In brief, what we sense as qualities are just these differences and the alterations that follow from them. In one note Nietzsche writes, 'N.B. Given particular alterations of *quantities*, what we sense as a variation in *quality* arises.'[58] Perception is primarily of alteration, since that is what matters to organisms living within a varying environment. The immediate content of sensation is determined by change, and the resulting quality enables a quick and apt response by the perceiver. What Nietzsche says about quantity and quality is consistent with this, but also points to the role of force and intensity in his epistemology.

Intensity is not built up from a collection of parts, each of which has its own being, but is grasped all at once. Yet it can be called a becoming because of its inner aspect of more and less. Intensity involves succession, and so is measured by ordinal rather than cardinal numbers. Gaston Bachelard argues that a simultaneous ordering constitutes a distinct, 'vertical' time, on the grounds that 'every order is a time'.[59] I would rather say: all succession is a becoming, but not necessarily in the form of time. Intensity is a *becoming without time.* Can it therefore provide the bridge that we are looking for? We have not yet found a sign language of sight and touch to make the variations in absolute becoming accessible to human knowledge. Nor have we completed a transition from becoming to time. This is really the same problem, but it seems that the concept of intensity is an important clue to a solution.

Nietzsche provides a further indication by introducing a theme central to the following discussion: the concept of *tempo.* He is explaining that there are several ways in which the illusion of being can arise: 'In a world of *becoming,* "reality" is always only a *simplification* for practical goals or a *deception* on the basis of crude organs, or a variation in the *tempo* of becoming.'[60] Are these really separate options, though? Our perceptual limitations are exactly simplifications for practical purposes, if not deliberate ones. The last case is a puzzle, if it is true that absolute becoming 'proceeds continuously with the same speed', the thesis of Herbart cited above. Spir's response was that absolute becoming provides no basis for the comparisons required to say that becoming proceeds faster at one time and slower at another, or always at the same speed. Yet changes undeniably *seem* to us to proceed at different rates, and sometimes we even observe things at rest and not undergoing change at all.

We also experience variations in tempo within ourselves. Nietzsche regards these as features of the life-process, influenced by environment and lifestyle, that have a determining influence on mental life and its expressions. His description of inspiration in *Ecce Homo* emphasizes fluctuations in tempo: 'now the pace quickens involuntarily, now it becomes slow'.[61] This passage is meant to describe the composition of the first part of *Thus Spoke Zarathustra* at the beginning of 1883, and it highlights the work's diversity of style and rhythm.

Its aphoristic passages represent the rapid tempo of thinking that Nietzsche singles out (usually at the expense of his fellow Germans) as that of the free spirit.[62] But he notes of other passages that 'the tempo of these speeches is a tender adagio'.[63] There is no inconsistency here, for either tempo may represent a strong and healthy physiology. On the other hand, either may be a symptom of decline. The Venetian writer Luigi Cornaro lived to an old age, Nietzsche tells us, not so much because of the restricted diet he advocated but because his metabolism was unusually slow. Nietzsche makes a similar claim for himself: 'My blood moves slowly.'[64] Yet a slower tempo of life, along with cooler senses and an earnest temperament – as found in the scientific mind and asceticism in general, he suggests – may represent 'a certain impoverishment of life'.[65] Meanwhile, those who feel a restless desire for constant activity, who 'fling themselves to the moment', are condemned by Zarathustra as trying to escape from themselves.[66] The strength of will that sees far ahead and spans long periods of time counts for more.

These illustrations (and more could be given) show how important the theme of variations in tempo is for Nietzsche's thinking about life and knowledge, and the relation between the two. But he is aware of the conceptual difficulties associated with these ideas since ancient times. The assumption that motion can be faster or slower is crucial to Zeno's 'Achilles' and 'racetrack' paradoxes. Are our concepts of space, time and causality adequate for describing things that move with different speeds? Nietzsche writes, 'The sense of time is given with the feeling of cause and effect, as an answer to the question about the degrees of speed in different causalities.'[67] This suggests a translation of degrees of intensity into differences in extensive magnitude, accessible to sight and touch. Yet the paradoxes challenge any such translation by implying the impossibility of comparisons between times or distances. After all, they ask, how can one infinitude be greater or less than another? When discussing Zeno in his Basel lectures, Nietzsche takes the ideality of space and time as providing a Kantian solution to all these puzzles. His later answer would use the language of construction and confabulation. We grasp intensities by translating them into the vocabulary of sight and touch. When one recalls how pressure gauges and thermometers indicate pressure and temperature as spatial distances, this is a familiar practice. Nietzsche, however, is proposing a much broader theory. He suggests that the purpose of our conception of time and space is to make intensities visible and thus graspable by our minds. As we shall see in the next chapter, his 'time-atom' theory draws on this idea to construct a whole system of natural philosophy.

Differences in speed bear on a fundamental question about time. Can there be a unit of time – that is, an interval that would be finite but indivisible, and therefore the least possible amount of time? Nietzsche seems to assume that no change could occur *within* such a unit of time: in other words, it would be a period of rest. The argument might be that any change would imply several distinct states (or locations, in the case of motion) in relation to this time, and that this either amounts to a contradiction or enables us to divide the time into two parts, contrary to the hypothesis. Thus, any proof of the impossibility of a state of rest is also an argument against a unit of time, and in turn against time as something in its own right. Nietzsche summarizes a standard argument for the infinite divisibility of both space and time, originating in Aristotle's *Physics*[68] but developed by later thinkers, and encountered in his reading of Spir.[69]

> The outer periphery of a wheel is constantly moving, just like the inner periphery, and although slower [*sic*], not at rest in comparison to the faster inner one. The difference between slower and faster movement is not to be determined in terms of 'time'. In absolute becoming force can never rest, never be non-force: a 'slow and fast motion of force' is not to be measured in some unit, for that is lacking.[70]

The hypothesis of extended but indivisible 'atoms' of time and space implies that motion can occur only at a single rate, by covering one unit of space in one unit of time. If anything appears to be moving faster than this, it must be jumping over some units of space as it proceeds along its way. If something appears to be moving more slowly, it must be staying in the same place for several units of time before moving on. An observer who fails to notice these jumps or pauses will have an impression of the object as being in constant motion, but that is an error. On this account, the belief that different processes occur at different speeds is an illusion, caused by our inability to perceive more than a fraction of the content of becoming.

How does the image of a rotating wheel support an argument against time-units? If a solid object is rotated, some of its parts will necessarily be moving faster than others. For example, points on the outer and inner peripheries of a wheel must be moving with different speeds, since they complete their revolutions in the same time but cover different distances along the way. Yet both are moving continually. It is not as if the faster one pauses from time to time to let the slower one catch up, or that the slower one skips over some parts of the distance in order to keep up with the other. If either of those things occurred, the solid wheel would break into pieces. On the time-unit model, however, there would have to be times when the outer point is moving and the inner one is not. How could the wheel remain rigid under those circumstances? When Nietzsche comments that the difference between slower and faster motion is not in terms of 'time' (the inverted commas are his) that probably means: in terms of the time-units that are implied in taking time as something in its own right. He concludes the note by repeating the point made at the beginning about our necessarily mistaken perception of becoming as bound up with persistence.

Although cryptic and even confused (notably in its careless assertion that the outer periphery of a wheel moves more slowly than the inner one) this text is a very useful one. It is designed to rebut any denial of differences in the tempo of becoming and, by the same token, any theory of finite but indivisible time-units. Differences in speed are the connecting link between the fluctuations of intensity in absolute becoming (which, although continuous, is not uniform) and the extensive magnitudes of time and space, which provide their conceptual expression. Nietzsche has said that variations in tempo are perceived as contrasts between motion and rest, and those contrasts are what create divisions between definite intervals or periods. Here another concept takes on an important role in the emergence of temporality: this is *rhythm*, defined by Plato as 'the order of motion'.[71]

Rhythm as the Form of Becoming

The theme of rhythm appears often in Nietzsche's early thought, where issues of becoming are framed in terms of the model of Greek culture presented in *The Birth of Tragedy*. He regards the awareness of time-relations expressed in poetic metre and musical rhythm as an essential part of the Greek experience of art and life. These conceptions are identified with the drive to impose form and limit on a world of contradiction for which Nietzsche uses the term 'Apollonian'. Rhythm is a patterning of time, and so a primary form of the world of appearance, as well as a central element in musical experience. It is not only about measure and limitation, but also about the creation of beautiful semblance. Nietzsche moves back and forth between these two aspects, enabling each to inform the other.

Music for Nietzsche in the Basel period is an expression of the will. Rhythm corresponds to what he calls the 'intermittence forms' of the will, dynamics to the quantity of the will, and harmony to the essence of the will.[72] Rhythm is not directly found in sensory perception – it cannot be identified with the syllables, tones or movements through which it comes into experience. Rather, it is their pattern, like the shape of an object, which is perceived when that thing is perceived but remains distinct from its material substance. Hence, rhythm is something of an abstraction: Nietzsche says that it is 'ideally present' as the measure or 'regulator' of what is actually seen or heard.[73]

Yet even abstracted rhythm is relatively close to the world of appearance. In contrast to this is harmony and also melody, which Nietzsche regards as an abbreviation for sequences of harmony.[74] In terms of Nietzsche's later epistemology, harmony can be seen as a paradigm case of *quality*, as he understands this important but elusive concept. It is essentially emergent, not from tones as such (since those can vary while a chord remains the same) but rather from the *differences* between tones in their intensive aspect – from their 'more' or less'. Whereas rhythm is a form of appearance, a construction of becoming as time, harmony is a becoming that is not a temporal phenomenon. This is why Nietzsche treats harmony (meaning both concord and discord) as the true essence of music, the source of its unique power to break through the veil of appearance and uncover the inner reality of the will.[75] Rhythm is a pursuit of individuation, but harmony stands for 'the pure essence of willing' that underlies all individual phenomena.[76] It is this disruptive power, exemplified for Nietzsche in the final act of *Tristan und Isolde*, breaking through the veil of illusion that constitutes the great achievement of music.

In *The Birth of Tragedy* Nietzsche postulates an early Greek experience of music, before the fateful encounter of Hellenic culture with the Dionysian cult, and formed along Apollonian lines:

> If music, as it would seem, had been known previously as an Apollonian art, it was so only as the wave beat of rhythm, whose formative power was developed for the representation of Apollonian states. The music of Apollo was Doric architectonics in tones, but in tones that were merely suggestive, such as those of the cithara. The very element which forms the essence of Dionysian music (and hence of music in general) is carefully excluded as un-Apollonian – namely, the emotional power of the tone, the uniform flow of the melody, and the utterly incomparable world of harmony.[77]

Can this purely Apollonian art be called music at all? It is an architecture in tones, Nietzsche explains, reversing Goethe's description of architecture as frozen music (*gefrorene Musik*).[78] In contrast, the Dionysus cult used sharp disruptions of time measure to undermine the Apollonian forms.[79] His Basel lectures on Greek rhythm approach the theme in far more detail. Coming from the direction of poetry, they do not overlap much with *The Birth of Tragedy*, yet the two together form a consistent picture. In the lectures Nietzsche sums up his model of rhythm in these words:

> Rhythm is an attempt at individuation. For rhythm to exist, there must be multiplicity and becoming. Here the quest for beauty reveals itself as the motive for individuation. Rhythm is the form of becoming, and in general the form of the world of appearance.[80]

When rhythm is described as the form of becoming, and meter (*Takt*) as 'the *form of time* itself',[81] the key word 'form' must be taken as indicating the operation of the Apollonian creative drive. Rhythm is the schema that turns becoming into time. Thus, time is understood from the start as an aesthetic phenomenon.

How does rhythm impose form? The answer seems to be: by providing the contrast between rest and change that is necessary to give beings like us any grasp of becoming. Nietzsche's account is largely based on Aristoxenus of Tarentum, a contemporary of Aristotle whose partially surviving textbook is the earliest available writing on the subject. Aristoxenus states that rhythm is not continual motion or rest, but an alternation of the two.

> Rest is indicated by the position or the note or the syllable. None of these can be perceived by the senses unless there is rest. And movement is indicated by the shift from position to position, from note to note, from syllable to syllable. The time-lengths occupied by the periods of rest can be recognised, but those taken up by the movements are too short to be recognised. They are like boundary-lines between the time-lengths occupied by the periods of rest.[82]

On this view, we perceive a succession of different states of affairs, but the change that occurs between one and another occurs in an instant, so that we do not perceive it but can only be aware that it has occurred from the outcome. Nietzsche comments: 'The time of the rest is perceivable by the senses, the time of the motion, of the transition, is unknowable.'[83] This is close to his descriptions of the human mind's inability to apprehend the actual process of becoming, and its substitution of the succession of states on either side, so that motion itself is inferred rather than directly apprehended, just as in the musical example.

Nietzsche's early writing on rhythm is quite extensive, and recent writers such as James I. Porter and Friederike Felicitas Günther have convincingly shown its importance for his thinking as a whole.[84] In addition to the Basel lectures, he sketched out contributions to scholarship on the subject. Several writers had argued for a rediscovery of the ancient Greek conception of rhythm without a misleading application of modern concepts. Nietzsche supports this approach: 'We must go back to Aristoxenus', he writes, echoing the neo-Kantian slogan 'Thus, we must go back to Kant.'[85] His account of the development of meter and rhythm in ancient thought depends on a distinction between two kinds of rhythm, one involving time-relations (that is, contrasting long with short sounds) and the other involving force-relations (contrasting loud with soft sounds).[86] The main change in the theory of rhythm, he says, is the replacement of the original Greek emphasis on time-proportions alone, found in Aristoxenus, by a model based on stress or emphasis – in fact, the same model that we still have today. Nietzsche suggests that the Greeks had a far subtler sense of these relations than we do, preoccupied as we are with stress and emphasis as the determining factor in rhythm. He later wrote to his musician friend Carl Fuchs: 'It is hardly possible for us to feel a purely quantitative rhythm, we are so accustomed to the affect-rhythm of strong and weak, of crescendo and diminuendo.'[87]

In 1875 Nietzsche revisited the Dionysian and Apollonian in a notebook entry which suggests a reinterpretation in line with his growing inclination towards a naturalistic outlook. The occasion was a reading of Eugen Dühring's *Der Werth des Lebens* and the surrounding text is, in fact, mainly a dialogue with that book. Dühring emphasizes the importance of periodic processes in both organic and inorganic nature. Often we are not aware of these, he observes: for example, we take light and sound to be constant impressions, although they are in fact rhythmic phenomena, consisting in wavelike pulses that are too rapid for our perception.[88] Nietzsche later speculates along similar lines that pleasure, usually taken simply as a particular feeling, may really be 'a kind of rhythm in the successive occurrence of weaker pains and their relations of degree, a stimulation by means of the rapid alternation of increasing and decreasing intensities'.[89] In this note, he follows Dühring in treating rhythm as a dynamic process, an interaction of forces giving rise to a patterning of time. He commends Dühring's use of an analogy between music and life, both taken as involving a mixture of 'harmony' and 'disharmony', and adds: 'incidentally, his theory is already contained, symbolically-mythologically, in my conception of the Dionysian and Apollonian'. He goes on to spell out this correspondence:

> The Dionysian is the disharmonious ground which longs after rhythm, beauty etc. The rhythm of organic life – how much does it adapt itself to the form of the incoming stimulus? First of all, the contradiction may be sensed, up to the complete annihilation of sensation, and on the other hand the rhythm of organic life can completely give in to the incoming rhythm, and go over to it, at least for a time – all this is the Dionysian phenomenon. In contrast to that is the measured attitude to the incoming stimulus, holding fast to one's own rhythm, the mutual co-ordination of two rhythmic formations, finally the transference of one's own rhythm to the incoming rhythm (= beauty) the Apollonian phenomenon.[90]

This is a naturalistic version of the contrast, centring on the role of rhythm in the life-process and the interactive relation between the organism and its environment. It accords with the emphasis on the physiological side of life that was eventually to come to the forefront of Nietzsche's philosophical thinking.

Any privileging of harmony over rhythm in understanding music's significance is revised dramatically in Nietzsche's later thought. From *Human, All-Too-Human* onward, one of his criticisms of Wagner is the domination of harmony at the expense of rhythm in his music. Wagner's 'endless melody', he says, 'sets out to break up all mathematical regularity of time and force and sometimes even to mock it; and he is over-abundant in the discovery of effects that to the older ear sound like rhythmic paradoxes and blasphemies'.[91] We should not take this as a comment simply on Wagner's music, for Nietzsche was aware of Wagner's writings as well. His 1870 essay on Beethoven warns against the musical effect of rhythm, understood as a regular or 'symmetrical' succession in time. Being derived from the world of everyday experience, Wagner argues, it can only distract attention from the inner world to which harmony alone gives access.[92] In an earlier work, 'Music of the Future', he asserts that harmony was 'completely unknown' in ancient music.[93] However, the Christian church disapproved of rhythm as closely tied to dancing, and hence to worldliness, and so developed harmony, 'which through its characteristic changes would from then on motivate the expression of melody as rhythm had formerly brought it about'.[94] Even so, rhythm was preserved in folk music and in time made a return: 'When, with the decline of the church, the demand for a secular application of music gained the upper hand with the Italians, the easiest solution was to restore to melody its original rhythmic quality and use it for song as it had earlier been used for dance.'[95] Wagner concludes with the emergence of Italian opera, emphasizing melody but minimal in harmony. He contrasts this with the more innovative developments of the German musical tradition leading from Bach to Beethoven – and, by implication, himself.

What Wagner wants, according to the later Nietzsche, is to avoid an architectonic crystallization or petrification (*Versteinerung*)[96] – in fact, what had been described as the Apollonian mode of music in *The Birth of Tragedy*. In consequence, he wrote to Carl Fuchs, the Wagnerian experience may be compared with swimming or floating, as distinct from walking or dancing.

> Rhythmic ambiguity, such that one no longer knows or is *meant* to know whether something is head or tail, is without doubt an artistic means with which wonderful effects can be reached: *Tristan* is abundant in it – , as symptom of an entire art, however, it is and remains the sign of dissolution. The part becomes dominant over the whole, the phrase over the melody, the moment over time (and the tempo too), *pathos* over *ethos* (character, style, or whatever it is to be called –), finally also *esprit* over 'meaning'.[97]

That description is the accusation made in *The Wagner Case* against Wagner, once praised by Nietzsche, at least in drafts, as having a 'gaze for long rhythms'[98] and an ability to maintain a grasp of long dramatic scenes in his operas. In Nietzsche's later writing the Wagnerian conception of an art of *pathos*, relying on harmony rather than melody or rhythm, is replaced by an emphasis on what he calls 'the great style', marked by a mastery of long periods and extended forms. Nietzsche complains that this is found in arts such as architecture, but not in the music of his own time.[99] At the same time, he claims something similar for his own literary achievement: 'The art of the great rhythm, the great style of long periods to express a tremendous up and down of sublime, of superhuman passion, was discovered only by me.'[100]

Contrasted to this stylistic mastery is the artistic phenomenon labelled as '*décadence*'. Rhythm and tempo are the outward expressions of inner conditions, Nietzsche says: it is the strong and healthy artist who is capable of working on the large scale while maintaining control of details, and not allowing them to take on an independent status. The weaker artist falls back on individual phrases and gestures as devices for producing calculated effects on the audience. Again, Wagner is the main target. Underlying these polemics, however, is a broader theory, positing a close link between structures of temporality and the physiological dimension of human life. In a notebook entry dating from 1885, Nietzsche explains that human perception involves the 'invention of forms', in accordance with our conditions of life:

> Man is a form- and rhythm-shaping creature; he is practiced in nothing better and it seems that he enjoys nothing better than the invention of forms. One need only observe how our eye immediately occupies itself, as soon as it has nothing more to see; it *creates* something to see. Our hearing probably behaves no differently in the same case: it *practices*. Without the transformation of the world into forms and rhythms there would be nothing 'the same' for us, and so nothing recurring, and so no possibility of experience and assimilation, of nutrition.[101]

Sameness is seen here as a product of form and rhythm, which in turn are created by the Apollonian drive towards order. Its outcome is the appearance of temporal limitations and divisions, and in turn of patterns and recurrences. Artistic expression is one important way in which this relation is realized, but not the only one. In the next two chapters we will return to the same theme from different directions. Chapter 3 will locate time patterns within scientific thinking, while Chapter 4 will look into Nietzsche's suggestion, based on contemporary biological theory, that the life-processes of living things provide measures of time that in turn determine their experience of the course of events. Despite these shifts of context, the concept of rhythm will continue to be an essential part of Nietzsche's approach to temporality.

Chapter 3

The Time-atom Theory

What would a purely Heraclitean view of time be? Nietzsche answers this by drawing on Schopenhauer, assuring us that 'As Heraclitus sees time, so does Schopenhauer.'[1] As Schopenhauer describes it, the form of existence in the empirical world is constant change, with no prospect of rest. We are like a man walking on a tightrope, who would fall if he hesitated even for a moment (an image that recurs in *Thus Spoke Zarathustra*). Further, in this world only the present moment has any reality. According to Schopenhauer, 'What has been, no longer is; it as little exists as that which has never been. But everything that is, is in the next moment already regarded as having been. And so the most insignificant present has over the most significant past the advantage of *reality*, whereby the former is related to the latter as something to nothing.'[2] He goes on to interpret this characterization as an argument for the *ideality* of time. We cannot help feeling 'in the very depths of our being' that this state of things is not the whole story, that there must be some deeper reality than the continual flow of becoming – a reality compared to which the world of space and time is only semblance.

For Nietzsche, such a feeling is not a profound insight but an illusion, or rather a rejection of the value of appearance and in turn a denial of life itself. Discounting it, he is left with a conclusion that for a true Heraclitean, the past and future have no reality. The present is the only genuine form of time. As for the parts of time, such as days or years, these too must be given up, since they extend beyond a single moment. Lacking either vocabulary, it now appears that the Heraclitean can have nothing to say about time. Aristotle reports that the most consistent advocate of universal becoming was Cratylus, 'who finally did not think it right to say anything but only moved his finger, and criticised Heraclitus for saying that it is impossible to step twice into the same river; for he thought one could not do it even once'.[3] Nietzsche takes this to be the fate of anyone who sees becoming everywhere: 'Such a person no longer believes in his own being, no longer believes in himself, sees everything flowing apart in moving points and loses himself in this stream of becoming.'[4]

The image of the flowing stream encourages us to identify absolute becoming with *continuity*. As a concept, this is as close as we can come to grasping something that is ungraspable by its nature. Yet continuity is itself difficult to conceptualize, as philosophers have found. The preference of the twentieth century was to reject any appeal to immediate intuitions of space or time, and attempt instead a purely mathematical construction of the concept. Mathematicians now define continuity in terms of the limit concept: a function is continuous at a point if and only if its limit as that point is approached indefinitely closely is the value at the point. But what does this have to do with the idea that comes to mind when we use the word 'continuous'? We are likely to be thinking of some process that continues over a period of time without interruption, such as an object's motion that may include variations in speed and direction, but not stopping and starting. Continuity at a point is not relevant to this image in any evident way, because we are thinking of the object not as located at any point but as moving along a path. To raise the question of its location in a particular place is to challenge the idea of motion. That is just the purpose of Zeno's paradox of the moving arrow. So the mathematician's concept seems to be not so much a formulation of our intuitive notion as a substitute for it.

Perhaps that is inevitable when points become the focus of attention. In an odd way, the doctrine of becoming turns into its opposite when the moment is seen as the sole reality. This chapter will look at Nietzsche's thinking on the same issue, which includes a startling claim that 'time is not at all a continuum'.[5] He was led to that conclusion in considering the possibility that time consists not of continuous intervals but of separate instants. This is what has come to be known as his 'time-atom' theory. Written in an 1873 notebook, Nietzsche's discussion of 'time-atoms' (or 'time-points') did not attract attention until it was discussed at length in Karl Schlechta and Anni Anders' exploration of his early philosophical thought.[6] The 'time-atom theory' label is theirs: Nietzsche uses the phrase at the end, but his heading is 'Motion in time'. Since then, scholars have looked closely at this speculative exercise, with its unexpected relations both to contemporary sources and to ideas about time and reality expressed by Nietzsche elsewhere.

These writers have approached the text from different directions. For example, Greg Whitlock emphasizes the link with Boscovich, while James I. Porter draws attention to parallels with the earlier lectures on ancient Greek rhythm and meter. Schlechta and Anders present the theory as informed by Nietzsche's reading of books borrowed from the Basel university library: not just Boscovich's *A Theory of Natural Philosophy*, but also Spir's *Denken und Wirklichkeit* and Zöllner's *Über die Natur der Cometen*. Each of these is relevant to some aspects of the theory, although not necessarily to others. There may be other sources, not identified and perhaps no longer traceable. In addition, Nietzsche's own contributions cannot be overlooked. The text is an exploration of a range of themes: space and time, cause and effect, the nature of material objects, and the relation of these concepts to the sensations on which our knowledge of the world is based. A good deal of interpretation is needed to make coherent sense of what are only sketchy notes.[7] My own reading is guided by a belief that behind the text is a broader purpose. To understand what Nietzsche is attempting – and why – we must look at this bigger picture.

The time-atom sketch synthesizes a number of themes which were running through his thinking at that time. It came in part from his engagement with early Greek thought and his wish to give an account of the philosophers of the 'tragic age' which would parallel his account of the origins of tragedy. He saw the philosophers as dealing with the same problem of knowledge and life, but raising the issue of a scientific understanding of the world by bringing all the phenomena of experience under some conceptual model. Nietzsche's treatment drew on his interest in scientific thinking, especially his reading of Boscovich and more popular writers on science, as well as of contemporary epistemology. In consequence, the context of his thinking shifted: it became a contribution to debates between contemporary writers such as Hermann von Helmholtz and the positivist school over the aims of scientific theory.

Nietzsche's attempted synthesis is undertaken in the service of a bold project: to construct an alternative to a scientific tradition going back to the ancient atomistic and Eleatic thinkers. His goal is nothing less than a Heraclitean science of nature. Achieving that, however, means turning a dominant scientific programme of his time on its head. This model takes modern physics to be about laws of *space*. In that conception, time places no essential role. For Nietzsche, the philosophical genealogy of this project is to be found in the tradition of Parmenides and the Eleatic school, which treated genuine knowledge as directed towards changeless being rather than becoming. The atomic doctrine of Leucippus and Democritus came out of this background and re-emerged in partnership with Newtonian physical theory to become the dominant scientific paradigm of the modern era. The most recent advances in physical science that Nietzsche knew about seemed to confirm this situation. The discovery of the conservation of energy made it possible to define physics as the science of force, equal in status with the well-established science of matter, chemistry. In both cases a principle of conservation was the basis for accounts of natural phenomena in which successive states of affairs were treated as simply rearrangements of the same amount of mass or energy. With the help of this principle physics achieved dramatic advances, overtaking chemistry as the most advanced and prestigious of the natural sciences.

This relation of quantitative equality leaves open the question: *Why* does one state give way to the other? Why not the other way around? Chemists spoke metaphorically of 'elective affinity', as if molecules preferred to be joined with one partner rather than another, and behaved accordingly by breaking off existing bonds and forming new ones. Nietzsche noted that the 'driving force' had been left out of chemical equations. 'We consider only results, and we posit them as equal in relation to content of force, we spare ourselves the question of the *causation* (*Verursachung*) of a change . . .'[8] Thus, the most ancient purpose of science, to search for the causes of phenomena, was abandoned. Some writers even asserted that the aim of science was not to explain but only to describe the world. If Nietzsche later seemed to adopt this view as his own, it was with an ulterior motive: to point out the need to locate the positivist model within an epistemology that would include a critical account of its self-imposed limits, and at the same time suggest new answers to the questions that had been deliberately set aside.

The Search for Ultimate Causes

The first point of reference for Nietzsche would have been Friedrich Zöllner's *Über die Natur der Cometen*, which he had recently been reading with admiration. In a speculative chapter entitled 'On the General Properties of Matter', Zöllner sets out to carry the idea of a science of spatial laws through to what he considers its ultimate form. The scientific model that provides his starting point is found in Helmholtz's classic 1847 paper 'On the Conservation of Force'. In this statement of the newly formulated scientific principle of conservation of energy, Helmholtz proposes what is, in effect, a comprehensive programme of scientific reductionism. All natural processes, he asserts, must be traced back to ultimate causes which consist in permanent, unchanging elements, and their discovery is the goal of scientific theory.

This conception recalls the Eleatic tradition, but Helmholtz's justification is drawn from a Kantian philosophy. He argues that if nature is to be intelligible (*begreiflich*) it must be assumed that every particular change has a sufficient cause. Where the causes that we identify are themselves changes, they must in turn have causes, and so we will proceed onward until causes are reached which do not involve change. 'The final aim of the theoretical natural sciences is thus to discover the ultimate and unchanging causes of processes in nature.'[9] The science of chemistry had been successful in finding unchanging causes, by postulating basic particles which never alter in their properties, only in their spatial combinations and rearrangements, and account for innumerable observed phenomena. Matter is thus assumed to be, as Helmholtz puts it, 'eternally unchangeable'. Can anything similar be said about force? Forces are functions of the spatial relation between material objects, Helmholtz states, and their effect is some alteration in this spatial relation – that is, motion. Using the findings of thermodynamics, he formulates a law of conservation of force by putting together what had been disparate concepts of *vis viva* (in today's terminology, kinetic energy), potential energy and heat. These forms may be transformed into one another, but the quantity of energy involved remains constant: that is, energy as such is neither created nor destroyed.

These are the ideas that Zöllner takes up. He agrees with Helmholtz on the authority of epistemology over science and endorses the thesis that scientific theory must trace all natural phenomena back to unchanging elements. Causality presupposes distinctions of time or space, or both together. Assuming atomism, the first class of differences is set aside 'through the assumption of forces that inhere in the interacting atoms with unchanging qualities from eternity to eternity'.[10] But Zöllner objects that Helmholtz's programme is 'one-sided', because its ultimate causes are constant in time, but not in space.[11] He insists that the need for causal intelligibility imposes a similar condition on the operation of forces in space.[12] This is more complex, because space may have more than one dimension. Only experience can tell us how many dimensions space has, as well as whether it is open or closed, and finite or infinite. Hence, various possibilities need to be covered. Arguing from the general principle that a similar spatial relation to a given object implies a similar force, Zöllner posits a universal law of force for any number of spatial dimensions.

This raises the question of action at a distance in space. Zöllner not only supports the idea but attributes it to Newton himself, despite contrary evidence. He is even prepared to consider action at a distance in time, quoting his Leipzig colleague, the physicist Carl Neumann:

> If one accepts (as occurs almost universally since Newton) that objects separated in space act on each other immediately, then it is just as legitimate to accept an immediate causal interaction between objects which are separated from each other in time: assuming naturally that such an assumption leads to consequences as fortunate as those of the first one.[13]

Zöllner's defence of action at a distance owes much to Boscovich, who could be seen as holding that *all* action is action at a difference. Citing both empirical evidence and metaphysical reasons, Boscovich posits a universal law of continuity stating that any quantity which changes from one magnitude to another must pass through all of the intermediate magnitudes of the same kind.[14] The problem now is that collisions are inconsistent with any such law. For when objects come into contact, their speeds are suddenly reduced to nothing at the moment of collision. But according to the law of continuity, a body cannot change velocity suddenly. Hence, bodies with different velocities cannot come into immediate contact. In fact, in Boscovich's view, distinct bodies can never come into contact at all. But since no part can come into contact with any other part, the primary elements of matter must be simple and unextended. Physical reality thus consists of mutually interacting points, governed by a complex law of force which involves both repulsion and attraction, according to spatial distance.

As Greg Whitlock observes, the dynamic world view arises 'historically and conceptually' out of the mechanistic world view.[15] The Boscovichian replacement of solid particles with unextended points of matter is an alternative theory within the same scientific programme. His centres of force, to use Nietzsche's preferred expression,[16] are just as permanent as Democritean atoms. The world must contain a certain finite number of them, on the Aristotelian assumption that an infinitude of coexisting things is inadmissible.[17] If points of matter could come into direct contact, their number might change, but this is impossible, since the force of mutual repulsion increases indefinitely at the smallest distances. Nietzsche was aware of this aspect of Boscovich's theory and criticized the 'dynamic atom' as still relying on the distinction between a subject and its activity, or a 'thing' and its effects. Points of matter are supposed to be in motion, he comments, but 'this always supposes, whether as the fiction of a little clump of atom or even as the abstraction of this, the dynamic atom, a thing that produces effects – i.e. we have not got away from the habit into which our senses and language seduce us'.[18] Thus, despite envisaging a shift in physical science from materialistic atomism to something like Boscovich's doctrine, Nietzsche demands a renewed attempt to break away from the traditional conception of permanent causes as the ultimate basis of explanation.

Nietzsche was plainly impressed by Zöllner's attempt to impose a single pattern of temporal and spatial constancy on force. The course of his thinking in the time-atom fragment is the same as Zöllner's. Both begin with reflections on time, space and force, and end with epistemologies of sensation. Yet Nietzsche's theory is not another version of the scientific theory set out by Helmholtz. It is a radical rejection of this paradigm, and that is what makes it of special interest. Nietzsche proposes a Heraclitean natural science in place of the Parmenidean model of changeless ultimate elements. Whether those are conceived as solid atoms or as Boscovich's points of matter, they represent a denial of becoming as a feature of essential reality. Nietzsche's intention is to turn this on its head by asserting the absolute impermanence of the basic elements of the world. His goal is amazingly ambitious: it is nothing less than to go back to Heraclitus and reconstruct natural science on the basis of his doctrine of becoming. This programme has several stages. First of all, it must attack the prevailing paradigm and show it to be untenable. Then it must indicate a viable alternative and show how the scientific knowledge that covers the range of natural phenomena can be reconstructed on this new basis. The time-atom sketch is designed to cover this entire ground within a few pages.

To some extent, it develops and brings together ideas that Nietzsche had already noted down about the themes of time, space and force, as well as a possible reduction of knowledge to sensation. One note made not long before the time-atom text links these reflections with the pre-Platonic thinkers of the 'tragic age', who according to Nietzsche variously employed art, logic, mathematics or asceticism in the service of 'the will'. He goes on:

> The will to characterise: its method to arrive at the rational. Essence of matter absolute logic. Time, space and causality as presupposition of *action* (*Wirkungs*voraussetzung).
>
> What remains are forces: in every smallest moment other forces: in the infinitely smallest interval always a new force, i.e. forces are not actual (*wirklich*) at all.

> There is no genuine action (*Wirkung*) of force on force: rather, in truth what exists is a semblance, an image. All of matter is just the outer aspect: in truth what lives and has effects is something quite different. But our senses are the *product of matter and things*, as is our *mind*. I think that one must proceed from the *natural sciences* to arrive at a *thing in itself*.
> The will is left over – when one subtracts the knowing intellect.[19]

In this 'logical' approach, time, space and causality are concepts assumed in order to account for action or effectiveness. That leaves action itself to be explained, and this is where force remains to be considered. Having set aside space and time (for the moment, anyway) we must think of force as timeless and spaceless: it cannot be something that occupies space or persists through time. In that case, however, one force cannot be understood as acting on another. Why not? The answer is that for a purely momentary force there is no 'next' force to act upon, since between that moment and any later one there must lie infinitely many others. The alternative notion of action of force at a temporal distance is not considered here, but left for the time-atom theory to explore.

All this, according to Nietzsche, means that force cannot be 'actual' (*wirklich*).[20] The unstated argument behind this claim can be found in *Philosophy in the Tragic Age of the Greeks*, where Heraclitus is identified with the proposition 'that the whole nature of actuality (*Wirklichkeit*) lies in its acts (*Wirken*) and that for it there exists no other sort of being'.[21] Nietzsche backs up this attribution by citing Schopenhauer's claim that matter and 'acting' (*Wirken*) are really the same thing. Only by acting on something does any material object occupy space and time, and only by acting on us is an object ever perceived. Hence, 'actuality' (*Wirklichkeit*) is a more appropriate synonym for materiality than 'reality' (*Realität*).[22] Schopenhauer's conclusion is that actuality is, like space and time, 'completely relative'. That could mean relative to the knowing subject, but Nietzsche takes it to show 'that each thing exists only through and for another one like it, i.e. existing only in the same fashion'.[23] He concludes that categories of time, space and force are all fictions, and draws on Schopenhauer again in the closing assertion what remains when one abstracts from all these creations of the intellect is 'will'.

Another note of late 1872 offers preliminary insights into the time-atom theory by addressing the relations between time, space and force, and introducing the notion of 'speed of causalities', soon to be redefined as 'motion in time'.

> Time space and causality are only *metaphors* for knowledge, with which we interpret things for ourselves. Stimulus and activity bound together: how this is, we do not know, we understand not one causality, but we have immediate experience of them. Every being acted upon (*Leiden*) calls forth an acting (*Thun*), every acting a being acted upon – this, the most general feeling, is

> already a *metaphor*. The perceived multiplicity, then, already presupposes time and space, in succession and coexistence. Coexistence (*Nebeneinander*) in time creates the sensation of space.
>
> Sensation of time given with the feeling of cause and effect, as answer to the question about the speed of different causalities.
>
> Sensation of space first to be derived by metaphor from sensation of time – or vice versa?
>
> Two *causalities localised* in coexistence – [24]

The idea here is that we interpret impressions made on the sense organs as the activities of those organs: for example, we say that the eye 'sees'. Time appears only in a derivative way, Nietzsche suggests, as an answer to the question about differences in 'speed'. How do we tell fast from slow forces? The answer must be: by how much time they take to produce their effects. This principle will be central to the time-atom theory. Space is even more derivative, a rationalization of perceived simultaneity: it answers the question about the distinctions between things taken as coexisting in time. One should also note in this discussion Nietzsche's striking scepticism about the status of sensations. They appear to be immediately given and self-evident in their content, but that is an illusion. Sensations are neither simple nor innocent: they arise out of hidden processes of metaphorical construction, and their seeming immediacy is only an effect on the perceiver. 'The simplest sensation is a boundlessly complicated story: there is no primal phenomenon (*Urphänomen*).'[25]

These ideas were in Nietzsche's mind when in early 1873 he borrowed a number of books on science and philosophy from the Basel university library. He wanted to find further resources to use for justifying the move to discard space and time, exploring the remaining concept of force in more detail and adding an epistemological dimension by linking this model with the concept of sensation, central to his thinking about knowledge of the world at this time. The outcome of this renewed study and reflection was the time-atom text, to which I now turn.

Time or Space?

Introducing the time-atom theory, Nietzsche explains his overall programme. He argues for the need to make a choice: between a science whose fundamental laws refer to space alone, setting time aside, and one based on 'laws of time' without any mention of space. He believes that the first option runs into an insoluble problem: it cannot explain how things at different locations can *act upon* one another. Hence, it must give up the notion of enduring things and in the end postulate a world that is timeless and changeless. This leaves Nietzsche to explore the second alternative. The puzzle now is to see how laws of space can ever come about from a scheme that originally contains no space. The text begins by setting up these dilemmas:

> Motion in time
>
> A B
>
> • •
>
> Space-point A acts on (*wirkt auf*) space-point B and vice versa.
>
> For that it requires some time, for every action has to cover a path.
>
> Successive time-points would collapse into each other.
>
> A no longer encounters with its activity (*Wirkung*) the B of the first moment. What does it mean that B still exists and A too still exists, if they do encounter each other?
>
> It would mean first of all that A is the same without change in both of the time-points. But in that case A is not an acting force, for an acting force can no longer be the same, since that would mean that it had not acted.
>
> If we take what acts (*das Wirkende*) in *time*, then what acts is something different in each smallest time-moment.
>
> That means: time proves the *absolute impermanence* of any force.
>
> All spatial laws are thus thought of as *timeless*, that is, must be simultaneous and immediate.
>
> The whole world at one stroke. But then there is no *movement*. Movement suffers from the contradiction that it is constructed according to laws of space and yet through the assumption of a time makes these laws impossible; i.e. it both exists and does not exist.
>
> Here the solution lies in the assumption that either time or space is = 0.[26]

The key concept here is 'acting upon' (*Wirken*). This is more fundamental than space or time since, as we have seen, it is identified with actuality by Schopenhauer and, following him, by Nietzsche. Hence, the task is to find out what concepts of space, time and force are needed to support it. Nietzsche starts with the proposition that a causal process is always a relation between points of space, and so must involve action at a distance. His further argument rests on two crucial assumptions. First, forces will always *take time* to cover a distance. Secondly, causal action must *make a difference* to what acts as well as to what is acted upon. Putting these together, he concludes that the unchanging causes demanded by Helmholtz and Zöllner cannot exist at all.

How reasonable are these premises? The first is already inconsistent with Newtonian mechanics. It is true that sound and light take time to cover distances, but the force of gravitation acts without lapse of time: if a mass changes its position, the effects determined by its gravitational attraction are undergone immediately by all other masses, even at vast distances. Nietzsche's second assumption may perhaps be seen as depending on this first one. If causal influence travels at a finite rate, it looks as if something is being *sent* from one object to another. In that case, some process must have occurred in the sender – in other words, the causal agent must itself undergo change. Having made these assumptions, Nietzsche can argue that no object really acts on another existing at the same time, since by the time the effect occurs, the source is no longer what it was. We know it is possible to see a distant star that has ceased to exist during the time taken by its light to reach us. For Nietzsche, this is not an unusual phenomenon, but the typical pattern of causal influence. Yet it constitutes a philosophical paradox. A spatial relation is between things that exist at the same time, and so it can correspond to a causal influence only if that is imagined as instantaneous, a concept that he has rejected.

Thus, he concludes, we are forced to choose between time and space. The world may be governed by laws of space, but must then be timeless. Alternatively, we can have a non-spatial world of elements located in time alone and related by laws of time. The second option is preferable because of the privileged status of time as a form of both inner and outer experience. Zöllner had speculated on the possibility of a sentient being having a 'spatial intuition of zero dimensions'.[27] Such a being might still be aware of successive states – for example, of varying heat and cold – and be able to correlate these with one another, but it would have no conception of material objects in the familiar sense. Zöllner uses this only as an illustration, but Nietzsche wants to go further into the idea and take it seriously.

On the face of things, the time-atom theory could hardly be more different from Nietzsche's other reflections on time. Instead of basing itself on a law of continuity, it postulates a complete discontinuity of time. These 'atoms' are indivisible elements, existing not in space but only in time. They are unextended: that is, each exists at a moment in time but has no endurance beyond that moment. It is this absolute impermanence, consistent with the doctrine of becoming, that enables the resulting theory to be regarded as a Heraclitean approach to natural philosophy.

Nietzsche now addresses the question of how any concept of an object or body can be arrived at on the basis of points in time. His answer depends on thinking of a point as capable of *repetition*.

> If I take space as infinitely small, then all distances between atoms become infinitely small, i.e. all pointlike atoms collapse together into one point.
>
> But since time is infinitely divisible, the whole world is possible purely as a time phenomenon, since I can occupy every time-point with the one space-point, and in that way posit it infinitely many times. One would therefore have to imagine as the essence of a body time-points in a distinct way, i.e. the one point posited at definite intervals. With each interval of time there is still room for infinite time-points: one could thus imagine a whole physical world, all supplied out of one point, but in such a way that we separate bodies into interrupted time-lines.

FIGURE 3.1 Time lines . . . composed of discontinuous time-points

Nietzsche draws a diagram in his notebook to show what several time lines, each composed of discontinuous time-points, would look like. This rough sketch is not easy to decipher. The representation that I give here is based on the photographic reproduction of Nietzsche's notebook pages provided by Schlechta and Anders in their book (see Figure 3.1).[28]

What one sees on this page is likely to be influenced by interpretation, and that is true of my version too. Yet understanding what is intended is the key to grasping Nietzsche's time-atom theory. The two columns are intended to represent what the text calls 'discontinuous timelines'. That is, the points in each column represent time-points occurring before and after one another. In each case, time is understood to be discontinuous, and to consist only in the unextended points, not in the intervals between them. So far, this is a common ground among interpretations. After that, disagreement begins.

What is the relation between the two columns? In a detailed analysis of the text, Greg Whitlock argues that they represent earlier and later segments of a single overall sequence of time-atoms.

> An attempt to see two and only two broken lines horizontally, I believe, is futile. Considered as columns, there seem to be two and only two broken lines. Flow of direction in time is irrelevant: what first and foremost concerns the reader is the meaning, if any, to the order and/or number of the points. For convenience sake, let me call the left hand column of six dots 'time line x.' A consideration of the relation between time line x and the right column indicates to me that the right column adds one to the number of points found across it in time line x, so I propose to call the right column 'right x + 1.'[29]

Whitlock adds that 'imagination takes x and x + 1 to be non-identical time-lines (and thus containing non-identical time points) by its very act of comparison'.[30] In this way, he concludes, distinct objects are imagined as existing 'side by side'.

This seems to me the right outcome, but reached through a mistaken line of thought. Let us start at the beginning. First of all, the diagram contains two columns, each forming a series of points at varying distances. These are the 'definite intervals of time' that Nietzsche has just mentioned. In contrast with Whitlock's analysis, I think that the *numbers* of points in groups within the two series is of no particular significance. It is noticeable that those who have copied the diagram from the notebook page have not agreed on how many points are on either side. Nor is there any reason to suppose that every one of the time-points in the second column occurs *after* all of the points in the first column. Why in that case place the two series side by side, given that earlier and later positions are already indicated in the vertical columns? Whitlock might well reply: in order to make a correlation between them. But I think it is important to realize that the points in one column are never *directly alongside* those in the other, so that any correlation is fictitious. At least, I believe that is Nietzsche's intention. When drawing the diagram, he probably alternated between the columns, not paying much attention to alignment: hence, the need for further explication.

It is instructive to compare this diagram with another one occurring later in the time-atom text. They are intended to tell the same story, as I shall explain later, and only their manner of presentation is different.[31] Even so, the first diagram can be read in its own terms. What needs to be kept in mind is that Nietzsche is talking about the way that bodies are apprehended by a being capable of representation. He is starting with the simplest case, where two bodies are imagined as existing at the same time. Each consists of a series of time-points (or rather, 'the one point posited at definite intervals'). These two series, however, have come about only through *separating* the points of a single series, the original unique succession of time-points. This is presumably what Nietzsche means when he says that 'we separate bodies into interrupted time-lines'. There now appears to be two series presented alongside each other. Since there is only one dimension of time, the two series must be thought as occurring in the same time. Now two important observations need to be made:

1. The points in one series cannot correspond directly to points in the other, since there could have been only one point in any position *before* the separation occurred. Thus, any printed version that shows points in one series as aligned horizontally with points in the other – that is, as occurring at the same moment – is misleading, despite any impression given by Nietzsche's very rough sketch, done only for his own use.[32] In that case, they would simply be the same point. In short, the two series are *concurrent*, but their elements are never *simultaneous*.

2. Nevertheless, we perceive points in one series as corresponding to points in the other. That is an error, caused by a perceptual apparatus which prevents us from noticing small time differences. Nietzsche says that we often perceive events as simultaneous where closer inspection reveals that they are not. 'Let us be mistrustful of all seeming 'simultaneity'!', he warns. 'There are bits of time fitting in there which can be called small only according to a great measure, e.g. our human measure of time.'[33] Because of such impressions of simultaneity, we see the two objects represented by these distinct series of time-points as existing at the same time. How then can they be told apart? The answer is, by being located at different places in space.

On this view, space is not something given in experience, as both empiricists and orthodox Kantians believe. It is a theoretical concept, required as a principle of individuation for concurrent processes that are taken as simultaneous existences. Space is what enables us to identify the objects that we have posited in order to account for recurring patterns in the relations between time-atoms. What Nietzsche's time-point diagram represents, then, is not just the separation of objects but the construction of space in imagination. He is making this point when he goes on:

> Now only a reproducing being is needed, one which retains earlier time-moments alongside present ones. In this way what we regard as bodies are imagined.
>
> There is no coexistence then except in presentation.
>
> All coexistence would be inferred and imagined. the laws of space would be wholly constructed and not establish the existence of space.
>
> The number and kind of the successive occurrences of that one posited point then constitutes the body.
>
> The reality of the world would consist then in one permanent point.
>
> Multiplicity would arise if there were conscious beings who could be aware of this point, repeated in the smallest time-moments: beings who took the point as not identical at different time-moments, and then took these points to be simultaneous.

The sudden introduction of a knowing observer is a disconcerting move, especially when this being is given the task of performing a synthesis of successive time-points. Such a capacity presupposes a temporal awareness that must be independent of the time-points undergoing this process of conscious reproduction. One can hardly avoid asking: has the continuous time so noticeably absent from his time-atoms simply been reintroduced at some transcendent level of reality? Or is the observing mind itself a part of the picture – that is, a constructed combination of time-atoms? Nietzsche does not tell us, and the theory's epistemological aspect remains undeveloped. Even so, it can be seen that these remarks agree with our interpretation of the diagram that they refer to. It is this conscious observer who constructs the two time lines alongside each other and takes them as coexisting bodies. There are earlier and later points within each time line, since each object is imagined as enduring through time. So the mind's activity involves both synthesis and analysis, together giving rise to a presentation of things in space and time.

The Order of the World

Let us look more closely at the process of construction that Nietzsche postulates. It assumes that the *intervals* between time-atoms display recurring patterns which can be recognized by an observer. The most basic learning here was done long before the invention of human language. Pattern recognition is the original epistemic achievement of all sentient beings, and it still underlies our most sophisticated claims to knowledge, as recent developments in cognitive science have shown. This capacity extends to distinguishing patterns even when they are presented at the same time. For example, when we hear orchestral music, our hearing performs a complex piece of data analysis: we can tell one instrument from another, and hear them as playing together, each one being located in a different place. The analysis that Nietzsche is proposing here is similar, except that the intended outcome is the empirical world, with all its objects in their relations to one another.

Nietzsche restates these themes and adds an epistemological dimension by suggesting that the time-atoms correspond to sensations, a theme that will recur at the end of the sketch.

> Translation of all laws of motion into time-proportions.
>
> The essence of sensation would consist in gradually sensing and measuring such time figures more and more finely; the mind construes them as a contiguity (*Nebeneinander*) and then explains the course of the world in accordance with this contiguity: a pure translation into another language, into that of becoming.
>
> The order of the world would be the lawlikeness of time-figures; but one would then have to imagine time however acting with a constant force, in accordance with which we could interpret only in terms of contiguity. *Actio in distans temporis punctum.* We have no means at all for setting out a time law by itself.
>
> We would then have a pointlike force, which would have a relation to every other later time-moment of its existence, i.e. whose forces would consist in those figures and relations. In every smallest moment force would have to be different; but the succession would be some or other proportions and the world as we know it would consist in the becoming visible of these force-proportions, i.e. their translation into spatial terms.

In atomistic physics one usually assumes atomic forces unchanging in time, thus *onta* in the Parmenidean sense. but these could not act.

Rather only absolutely changeable forces could act, ones that are not the same at any moment.

All forces are only functions of time.

1) An action between successive time-moments is impossible: for two such time-moments would collapse into one another.

Thus every action is *actio in distans*, i.e. by leaps.

2) How an action of this kind *in distans* is possible we do not know at all.
3) Fast, slow etc. in [*sic*] the whole nature of this action.[34] I.e. forces, as functions of time, are expressed in relations to nearer or more distant time-points, in other words as fast or slow. The force consists in the degree of speed. The highest speed of all would consist in the action of one time-moment on the nearest, i.e. it would then = infinitely great.

The greater the slowness, the greater the intervals of time, the greater the distance. Thus the relation between distant time-points is slowness: all slowness is naturally relative.

The patterns that Nietzsche calls 'time-figures' are central to this line of thought. Time-proportions are a familiar concept not only in musical theory but also in physical science. Galileo's law of falling objects states that the distances travelled are proportional to the squares of the elapsed time, so that an object moving under the influence of gravity will take progressively shorter times to cover the same distance.[35] If these successive stages are arranged along a line, the 'leaps' will become smaller and smaller. For Nietzsche, however, such a sequence does not represent various actions of the *same* force. On his assumptions, a force acts just once, and in acting it ceases to exist, for its sole actuality consists in its acting (*Wirken*). The image of time-proportions may be contrasted with a spatial law of force. In a Newtonian account of the same phenomenon, the law of falling objects is derived from an inverse square law of force. This is a purely spatial law which is applicable at any given instant to objects existing at that time. A more complex law of force is presented by Boscovich, involving attraction at some distances and repulsion at others, but the same comment applies. His law too indicates a relationship between force and space.

What about 'motion in time'? Nietzsche characterizes forces as 'fast' or 'slow', and explains that a fast force is one that acts at a short distance in time, and a slow force one that acts at a great distance in time.[36] These terms are 'naturally relative', he adds, since there is no independent measure for them. Nietzsche later favoured a Humean account of causality, but at this stage he is thinking in terms of Schopenhauer's principle that *acting* is the most fundamental property of things, not derived from anything further. With space out of the picture, the only way of characterizing the nature of this acting is in terms of time itself.

Nothing more is said about 'fast' and 'slow'. However, speaking of the 'speed' of a force suggests an intensive magnitude, like that of an object moving in space. Further, this conceptualization leads on to the subsequent discussion of sensation. As we noted earlier, sensations have intensity, and are our evidence for the presence of intensive magnitudes in external things. Further on, Nietzsche writes, 'The dynamic time-point is identical with the sensation point.' The discussion of absolute becoming in Chapter 2 located the origins of sensation in fluctuations of intensity and consequent interactions of force, in which quality emerges from intensive differences. Now we see Nietzsche putting this model to use.

A different interpretation of 'speed' in the time-atom sketch is given by Whitlock. He thinks that Nietzsche is talking about acceleration in the literal sense of non-uniform spatial motion, and so invokes a comparison with Boscovich's spatial law of force, concluding that 'Nietzsche has all but drawn a diagram of the "Boscovich curve".'[37] This assimilation implies a direct relation between Nietzsche's time-figures and Boscovich's spatial law of force. One needs to remember, though, that the sketch's heading is 'Motion in time'. Space was explicitly eliminated from the picture at an early stage, pending some reconstruction which is not yet in view. So reverting to the familiar concept of motion, and accordingly seeing spatial relations in the diagram, is premature at least. When Nietzsche says that the greatest speed would imply one moment's acting on 'the nearest', he means the nearest in time, not in space. Further, 'nearest' is relative here, since on the assumption of infinite divisibility there can be no single 'nearest' moment, only an infinitely close approximation of moments. This is really a way of saying: the greater the speed, the nearer the moment acted upon.

We can now consider the text's second diagram of time-proportions (Figure 3.2), which is even more instructive than the first diagram, but needs some careful attention.

The diagram is intended to convey the process of analysis that gives rise to a multiplicity of objects out of a single sequence of time-atoms. So was the first one, but what is intended is now easier to see. At first sight, it looks as if Nietzsche has crossed out a vertical line, to signal that time is not a continuum but only a collection of distinct points. These are presumably infinitely many in number, since between any two points others can always be found. It is only for convenience that a limited number are given in the diagram. On the other hand, one can argue that since there was no line to cancel the diagonal lines may be there simply to emphasize the separateness of the points – by keeping them apart, so to speak.

The set of points is supplemented by curved lines, crossing one another as they link separated points. On looking carefully, one sees that they indicate three distinct sequences of points. It was two such sequences that were represented in the earlier diagram. Motion is continuous in Boscovich's sense if and only if a thing moving from one place to another passes through all the

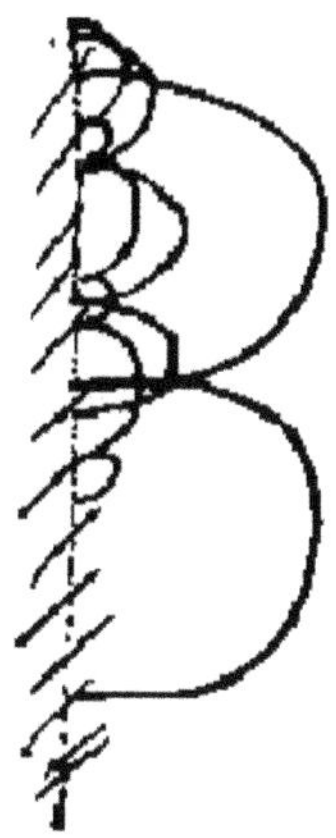

Time line.
Real: one space-point.

Relations of its various time states.

What the relations consist in.

No movement in time is continuous

We measure time by something permanent in space, and therefore we assume that between time-point A and time-point B there is some continuous time. But time is not at all a continuum; rather there are only totally distinct time points, not a line. Actio in distans.

One can speak only of time-points, no longer of time.
The time-point acts on another time-point, hence dynamic properties are to be assumed.
Time-atom theory.

FIGURE 3.2 The process of analysis that gives rise to a multiplicity of objects out of a single sequence of time-atoms

intermediate positions.[38] These movements from one time-point to another do *not* pass through all the time-points in between, and that is why they are represented by curved lines, not by straight lines joining the points. As before, the correlations that constitute each sequence are the contribution of the conscious observer, whose power of retention creates links between one time-position and another by taking points to be 'the same' as earlier points. In his commentary on the diagram, Nietzsche raises the question: what do these theoretical constructions consist in? He thinks that the answer has to do with the dynamic properties of time-points. That is to say, these are *forces* acting on one another.

With this in mind, we can now say what 'motion in time' means. Nietzsche's diagram is intended to show a variety of fast and slow forces at work, the slow ones being those making the biggest leaps. The parallel with fast or slow movement in space is metaphorical: a force does not move in time in the way that an object moves in space, by being wholly present in one place and then in another. Thus, a given force is fast or slow by its nature, and does not change in that respect. Further, these forces clearly work by action at a distance in time. In Chapter 2 we noted an objection to any model of cause and effect as separated by time. How can it be guaranteed that the effect will follow the cause, given the possibility that other factors will intervene during the interval? Nietzsche avoids that threat by keeping each force separate from all others. There is no conflict,

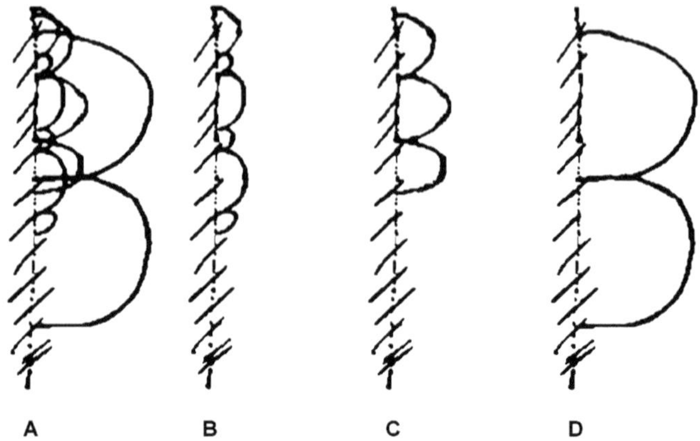

FIGURE 3.3 Combination of repeating patterns

no entanglement or 'knotting', for the curved lines do not interact with one another.

Nietzsche's diagram may seem to be an irregular collection of time-relations. In fact, it is a combination of repeating patterns, as this illustration should make clear (see Figure 3.3).

Here is Nietzsche's notebook sketch, labelled A, along with three further diagrams, labelled B, C and D. These are just the three constituents of A. (I have listed them from the top down, but it is possible that Nietzsche drew them in a different order.) It can be seen that they are all regular figures: B alternates long and short intervals, while C and D repeat the same intervals. Nietzsche's diagram is essentially the same as his previous one. The difference is that instead of being placed side by side, the concurrent sequences of points (now three rather than two) remain within the original time line and are indicated by the added curved lines. Those lines are only a pictorial device to indicate the grouping of points on the time line. Hence, they could be omitted from B, C and D, and then we would just have three sequences of time-points, looking very much like the two sequences of Nietzsche's first diagram. But the curved lines are helpful in conveying how the sequences represent a conception of time as discontinuous. This fact is particularly clear when the three curves are combined, since we can then see the time-points that are passed over by any given curve.

In *Ecce Homo* Nietzsche speaks of artistic inspiration as involving 'an instinct for rhythmic relationships that arches over (*überspannt*) wide spaces of forms – length, the need for a rhythm with wide arches, is almost the measure of the force of inspiration, a kind of compensation for its pressure and tension (*Spannung*)'.[39] Even though this passage was written much later, it throws light on the time-atom diagram. One could say that B, C and D represent three distinct rhythms, like the concurrent meters that Aristoxenus says are needed for rhythm in the true sense.[40] The strongest force is the one seen in D, with its wide arches symbolizing long periods. Nietzsche's theory is that these curves represent the patterns constructed by the observer capable of memory and association, and interpreted as indicating distinct things, existing alongside one another through the same time period. Thus, the diagram as a whole represents what appears in conscious experience as three coexisting objects.

Nietzsche raises a problematical issue for his own model when he mentions *measuring* time. As he points out, measurement of time requires something to be in space, for example, an object moving (or rotating) at what is assumed to be a constant rate. If time were to consist of finite but indivisible units – that is, of 'time-atoms' in the sense of *chronoi protoi* – then measuring duration would simply be a matter of applying this standard. But Nietzsche's atoms are unextended and, as he indicates, there can be no question of their aggregation. Between any two points lie an infinitude of other points, providing no basis for comparisons of magnitude. It is finitude, not infinitude, that raises the unanswered questions here.

A related issue is raised by J. F. Herbart, who discusses the idea that time consists of time-points and states an objection to this theory: 'We allow time to fall apart into *time-points,* but then time will not allow itself to be put together again from those points; hence, we insert time *between* the points, as if there could be time where there is no time-point.'[41] Conversely, Herbart argues, if time is identified with these points, then nothing can *occur* between them, since there is no time there. Rather, it must be assumed that 'change itself falls *within* the time-point'.[42] To respond adequately to these objections, Nietzsche would have to posit something like Kant's transcendental imagination to account for the construction of 'wide spaces'. One could argue that this is implicit in the early notes on epistemology, especially in his frequent appeal to notions such as metaphor and projection. Neither of these figures within the time-atom text, but it is intended only to state the overall programme, leaving the details to be spelt out elsewhere.

Nietzsche concludes his sketch with a brief review of the argument that gave rise to his time-atom theory. There are three main stages of reduction in this line of thought. The first is supplied by physical science, or rather by Boscovich's theory of points of matter. The second stage depends on Nietzsche's own argument for the need to choose between a negation of time and a negation of space, and the reasons that he gives for preferring the second option. The third stage is the sensationalistic account of perception found not only in Zöllner and Spir but in positivist philosophers of science such as Ernst Mach. Putting all these together, Nietzsche gives a summary of his theory:

> It is possible, then:
> 1) to trace the world as we know it back to a theory of pointlike space-atoms.
> 2) to trace this again back to a theory of time-atoms.

3) the theory of time-atoms comes together with a theory of sensation. The dynamic time-point is identical with the sensation point. For there is no simultaneity of sensation.[43]

This 'theory of sensation' is not present in the time-atom text, where sensation is mentioned only in passing. What Nietzsche has in mind can be seen in Zöllner's treatment of this theme in *Über die Natur der Cometen.* Zöllner too wants to reduce physical theory to a theory of sensation. He argues that sensation is a much more fundamental fact of consciousness than any awareness of space, motion or causality and argues, in the same way as Lange, that no sensation can ever be explained from the properties of matter.[44] His account becomes particularly unorthodox when he speculates on a universality of sensations of pleasure and pain throughout nature, even in inorganic phenomena.[45] Nature, he concludes, always acts with the conscious or unconscious aim of reducing sensations of unpleasure to the minimum.

Nietzsche echoes some of these ideas elsewhere, but gives only a few indications here. He identifies sensation with the point of force, but also says that 'The essence of sensation would consist in gradually sensing and measuring such time figures more and more finely.' This is a lot of work for sensation to do, but memory and imagination are also involved. The 'reproducing being' has to retain earlier points in order to correlate them with later ones, and to construct a space within which they can exist as 'bodies' alongside other bodies. The full extent of Nietzsche's scepticism does not appear in the time-atom sketch, but comes out in other notes of the same period.[46] Even the fundamental concept of force is a fiction, he argues there. All we are given in experience is sensations, and these are neither spatial nor temporal. So the construction of time can occur only on the basis of experiences for which there is little or no literal vocabulary, allowing a great deal of scope for interpretation. The basic principles are clear enough. Space, time and force are fictional constructions; our sensations and representations are given, but underlying them is the will, which remains 'when one subtracts the knowing intellect'.[47] It is the will's fluctuations and intermittences that provide the basis for succession and rhythm, the forms of appearance that turn becoming into time.

Beyond the Time-atom Theory

Comparing the time-atom theory with Nietzsche's later thinking about time, one is struck by the contrasts. Here he asserts that time is not a continuum, and that causality is action at a distance in time. The denial of continuous time is in contrast with Boscovich's essentially Aristotelian approach, in which time consists of parts of time, and instants are only the boundaries between these, not themselves parts of time. Later Nietzsche returns to a strong commitment to continuity, as a corollary of the doctrine of becoming. He now denies that cause and effect can be located apart in time. 'Two successive states: the one cause, the other effect, is false. The first state has nothing to effect, the second has been effected by nothing.'[48] In fact, he abandons the whole idea of causality as a relation between two distinct events or states of affairs. It is replaced by the concept of a process (*Geschehen*) that takes time, and out of which cause and effect are abstracted by us in a misleading fashion. Hence, the dilemma to which the time-atom theory is one solution, the need to choose between changing and unchanging forces, is no longer encountered.

Nevertheless, the programme of the time-atom theory – the formation of a Heraclitean alternative to the scientific hegemony of Eleatic atomism – is still Nietzsche's goal in these new speculations. He draws on the vocabulary of thermodynamics, speaking of energy continually accumulated and discharged, neither created nor destroyed but transforming itself while remaining the same. Further, a continuing allegiance to Boscovich is evident in his explorations of the 1880s. However, I do not think this was at the forefront of his mind. Boscovich figured as a standby in his critical engagements with philosophical materialism, to be invoked as a counterweight to the prevailing model of atomism.[49] He would have been relying on the reading of nearly ten years before, and on the Boscovichian content of works such as J. G. Vogt's *Die Kraft*, which was certainly studied in 1881, the year of the thought of eternal return.[50]

Nietzsche's later commitment to continuity is accompanied by an opposition to atomism in all its forms, including the dynamic version. He attacks the 'atomistic need' (an echo of Schopenhauer's 'metaphysical need') as a human weakness to be explained – or explained away – by scientific psychology.[51] In his rival model, the basic elements of reality are temporally extended processes. These not only overlap in time but are held together like the intertwining of strands in a rope. The account that Nietzsche opposes is one of causal relations between distinct momentary events or states of affairs occurring at different times. But is not the time-atom theory just the most consistent version of such a model? This brings us back to action at a temporal distance. In several later notebook entries, Nietzsche expresses himself against any such notion.[52] But it would be more accurate to say that his conception of process dissolves the whole issue. A process is not like a 'centre of force', where a field surrounds some unextended point. It is extended by its nature, and this is what makes possible the interactions with other processes for which Nietzsche uses metaphors of 'intertwining' and 'knotting'.

What about action at a distance in general? Without the mutual interaction specified by Newton's third law of motion, action at a distance becomes an intriguing and mysterious concept, and yet Nietzsche is willing to take it seriously. In the second of the *Untimely Meditations*, written about six months after the 'time-atom' sketch, he attacks the 'historical critic' who sets out to investigate the origins of Christianity, or the lives of great composers, in a way that replaces a living influence with detached, objective knowledge. The effect, he says, is 'to make every spiritual *actio in distans* (action at a distance) impossible'.[53] The idea is that the great cultural models of past ages are capable of acting *at a temporal distance* on the present day. This seems to imply the power to act upon something without being changed – and in particular, without being affected by that object. It is the notion of an unmoved mover, which occurs in Nietzsche's more cryptic speculations on the world as 'following from' a guiding spirit rather than the power of a driving will.[54] In some places he associates action at a distance with the power of the feminine, which attracts without itself being attracted. 'The magic and the most powerful effect of women is, to use the language of philosophers, an action at a distance, an actio in distans; but this requires first of all and above all – *distance*!'[55]

Further, Nietzsche turns action at a distance into the model for *all* causality in one notebook entry dating from 1885:

> – that it is the will to power that also directs the inorganic world, or rather, that there is no inorganic world. 'Action at a distance' is not to be set aside here: something attracts something else, something feels itself attracted. This is the fundamental fact: in contrast, the mechanistic idea of pressure and impact is only a hypothesis on the basis of visual appearance and touch, even if it is regarded by us as a regulative hypothesis for a world of visual appearance![56]

Of interest here is the identification of action at a distance, on the one hand with life (or 'the organic'), and on the other with sensation, which provides the 'fundamental facts' on which any conception of the world is based. Despite Nietzsche's appeal to the will to power, the note as a whole is primarily about the nature of the organic world. Every living thing, he suggests, interprets and evaluates its outer world in accordance with its specific conditions of existence. For beings like us, the projected outcome is a world constructed by the vocabulary of visual appearance – that is, a world of stable, identical objects moving and interacting in space. Yet action at a distance, even in the inorganic world, can be seen as presupposing a kind of mutual cognition (*Erkenntnis*).[57]

What was wrong with the time theory? Why did it disappear from Nietzsche's thinking in its original version? Nietzsche's close relationship with Paul Rée, beginning in 1875 and lasting for six turbulent years, led him to adopt a far more positivistic approach to natural science, including a Humean approach to causality. Perhaps he also realized that his attempt to create a Heraclitean science had led to a hyper-Eleatic apotheosis of the moment as a radically isolated monad, without cause or effect, and that this was unproductive for any understanding of the world. In *Philosophy in the Tragic Age of the Greeks* Zeno is imagined as criticizing the ancient atomists for their talk of interactions between their atoms. In Nietzsche's later theory of becoming as consisting in entwined processes, the moment is an unreal abstraction from becoming. But, one may say, what about that notable 'moment', the one that features in one of the central chapters in Part Three of his most profound book, *Thus Spoke Zarathustra*? That is one of the complexities of this key text to be unravelled in the following chapters.

Chapter 4

Motion, Ways and Time

The discussion so far has focused largely on Nietzsche's early work. That is not because his thinking about time and becoming is found only there: on the contrary, many of his most important considerations appear from the time of *Thus Spoke Zarathustra* onwards. How far do these later ideas differ from the earlier ones? Although it suited Nietzsche to divide his writing career into stages, a practice copied by his commentators, one can find many continuities in his thinking. We have already seen an example, in the relation between his studies of musical rhythm and the central role of 'time-proportions' in the quasi-scientific 'time-atom' theory. This same theme will appear in another form later in this chapter, when we see how regular periods of time (such as days or years) are accounted for in *Thus Spoke Zarathustra*. Still, it is also true that new themes appear in Nietzsche's later work, including those for which he is best known, such as the death of God, nihilism, the *Übermensch* and the will to power. Most notable among these for its bearing on temporality is the thought of eternal return, central to *Thus Spoke Zarathustra* and repeatedly asserted by Nietzsche to have radical implications for human existence.

What distinguishes Nietzsche's later treatments of temporality is not just new themes, but a shift to new forms of expression. Why did this come about? In part, as a more adequate reflection of his intellectual temperament, never fully satisfied by the cool, aphoristic style that he had favoured during his philosophical partnership with Paul Rée. 'New ways I go,' Zarathustra says, 'a new speech comes to me.'[1] These new ways (of which we will soon hear more) cannot be travelled using the 'worn soles' of old tongues, which have proven too slow for him. Instead he wants a medium to communicate thoughts that are capable of transforming the lives of those who 'incorporate' them. Most evident in the sequel is a systematic use of figurative and poetic language. This is not simply for the sake of literary effect: an advantage of metaphor is that it gives access to conceptions that would require a complex philosophical vocabulary, or even be impossible to convey in literal terms.

In the central parts of *Thus Spoke Zarathustra*, a particular scheme of metaphor runs through the chapters in which temporality is a main theme. These are metaphors of *motion*. Since motion is the most basic symbol of becoming, as we noted earlier, it is not surprising that it is also a metaphor for time. Ordinary language confirms this. We commonly speak of time in terms of motion: future events are coming, they arrive and depart, one looks ahead to the future and back to the past, and so on. These metaphors are elaborated in *Thus Spoke Zarathustra.* First of all, Zarathustra speaks of different *kinds* of motion, and sets out what they are: standing (the privative case, so to speak), walking, running, climbing, dancing and flying. Since motion is taken as being along a 'way' of some kind, there must be different ways corresponding to these kinds of motion. For example, running takes place on a 'lane', while climbing is on a 'mountain path', in Nietzsche's terminology. Further, as Zarathustra's listing indicates, a sequence is involved here. It is what Nietzsche calls an 'order of rank' (*Rangordnung*), implying that one kind of motion can be said to embody more power and freedom than another. That in turn points to broader contrasts between general forms of life, often expressed by Nietzsche in terms of personality types such as 'hero', or of symbols such as 'lion'. Further still, the concept of a form of life lends itself to presentation in and through narrative. Within *Thus Spoke Zarathustra* these narratives surround Zarathustra's discourses and also figure within many of them. In the following discussion we will see how they are used by Nietzsche to develop and extend his metaphors of temporality.

Zarathustra's 'ways' are not always easy to decipher. Only in one case is a ready-made interpretation provided for readers. In the chapter 'On the Vision and Riddle' Zarathustra describes a gateway that stands between two endless lanes and has a name: 'Moment'. This word has a temporal meaning: it stands for the present, as distinct from past and future. Hence, the image is clearly intended to symbolize the forms of time. My discussion will start, however, at the beginning of that chapter, which contains a different way, one that represents not the past or future, but rather finite periods of time such as days or years.

Zarathustra and the Spirit of Gravity

In 'On the Vision and Riddle', Zarathustra is addressing the sailors of the ship on which he is travelling. He calls his hearers 'bold searchers and researchers', and praises them as seekers after knowledge who are willing to venture on dangerous travel into uncharted regions. As he puts it, 'you do not want to grope along a thread with cowardly hand; and where you can *guess*, you hate to *deduce*'. Since Zarathustra is, in effect, travelling on a passenger service ('they often took along people who wanted to cross the sea') this portrayal seems excessive, even if the ship is out of sight of land for several days. Still, his general point is an important one. The 'thread' is an allusion to the legend of Theseus, who relied on a self-winding thread to find his way out of the labyrinth after his encounter with the Minotaur. A thinking that deduces is one that follows a thread of argument to its conclusion. It avoids the danger of error by relying on the security of logical reasoning. But danger is just what the heroic thinker welcomes, and that is why such a person is not a dialectician but a lover of riddles.

Riddles are intellectual puzzles for whose solution there is no methodical procedure. Neither factual knowledge nor skill in demonstrative argument are of much use here. Riddles confront the hearer with a challenge, all the greater when a wrong answer is liable to incur a severe penalty. Yet some are attracted to riddles just because of the dangers associated with them, which require, above all, courage on the part of the respondent.[2] All these points are relevant to the discourse that follows: they provide directions for use, so to speak.

Zarathustra does not present the vision and riddle to this audience straight away, but prepares for them with a narration which is important in its own right. He describes an encounter between himself and his 'arch-enemy', the spirit of gravity. This figure has already been mentioned in *Thus Spoke Zarathustra.* In Part One Zarathustra says, 'I would believe only in a god who could dance. And when I saw my devil I found him serious, thorough, profound, and solemn: it was the spirit of gravity – through him all things fall. Not by wrath does one kill but by laughter. Come, let us kill the spirit of gravity!'[3] Later in Part Three, the chapter 'The Spirit of Gravity' gives an extended treatment of this symbolic figure. But he appears in person only in 'On the Vision and Riddle'. At that stage, the spirit of gravity has been identified as standing for a set of feelings and attitudes – and, it is implied, values – in direct opposition to the ones that Zarathustra champions. He is serious and earnest, and will later be explicitly linked with the burden of moral responsibility described in Zarathustra's first discourse, 'On the Three Metamorphoses'.

Many writers take this narrative to be the vision that Zarathustra has told his listeners will be revealed to them alone.[4] In fact, it is a report of an episode that Zarathustra says occurred 'not long ago'. Whether he is telling the truth in this is another matter, of course. Still, however unusual the events he describes, and however symbolic their function, the story itself is a factual account. It begins with Zarathustra climbing on a rough and steep mountain path, in the gloom of twilight. Sitting on his shoulder is a strange creature, described as half dwarf and half mole. This is the spirit of gravity, the embodiment of those forces that impose both physical and moral constraints upon human beings. Being lame, the dwarf is using Zarathustra as a means for reaching the heights, yet he not only weighs Zarathustra down but also engages in pouring 'leaden thoughts' (*Bleitropfen-Gedanken*) into his ear, reminding him that whatever goes up must go down, and that the stone thrown high may fall back on the philosopher and crush him. Zarathustra is sorely oppressed by both physical and mental burdens.

Commentators have made desperate efforts to interpret this passage. Some suggest that the 'leaden drops' are an allusion to the *modus operandi* of the fratricidal Claudius in Shakespeare's *Hamlet.*[5] The unstated implication that Zarathustra and the spirit of gravity are siblings is intriguing, but there are too many discrepancies. The dwarf is not using henbane or any other poison.[6] Nor is he choosing the victim's ear to avoid any signs of murder (and thus 'abuse the whole ear of Denmark') since it does not appear that he intends to kill

Zarathustra, at least in a direct way. Nor is this method designed to catch his victim unaware, for the 'leaden drops' are the words the dwarf is whispering and the thoughts that they embody. Zarathustra may be in some dreamlike state, but he is well aware of what the dwarf is saying to him.

Another reading of the leaden drops refers to a 'mediaeval torture', apparently involving dropping molten lead into the victim's ear.[7] Even given such a practice, and taking it to be a means of torture, despite the prospect of swift death, it would not fit the case well. For one thing, in Nietzsche's drafts the dwarf's breath is described as 'ice-cold'.[8] More importantly, the dwarf is not putting Zarathustra to the question. He utters his scornful remarks, designed only to discourage, and falls silent. When questions are eventually asked, it is Zarathustra who asks them. In the text, the dwarf is described as making Zarathustra 'lame' – that is, as weighing him down, and so making his climbing more difficult and painful. This weighing down is all that an interpretation of the 'leaden drops' has to explain. Any addition of further details (a surprisingly frequent practice among Nietzsche scholars) makes the resulting reading less rather than more plausible.

A far more apt comparison is with Mark Twain's story 'The Celebrated Jumping Frog of Calaveras County', which Nietzsche had read aloud to his Basel circle of friends, ten years earlier, from his copy of the German edition.[9] Twain tells how the successful career of the champion jumper of the title – a frog named Dan'l Webster – is brought to an end by an act of sabotage. A visiting 'Yankee' accepts the owner's challenge to wager on the frog's performance against a rival contestant, and wins by an underhand device. He secretly weighs the frog down by forcing him to swallow lead shot. The frog is unable to jump, and the stranger makes his departure before the trick is discovered. This story supplies all that is needed to account for Nietzsche's text.

The thoughts themselves are, appropriately enough, all about falling. 'You threw yourself up high, but every stone that is thrown must fall', the dwarf says, and repeats the line with a variation: however far it has been thrown, the stone must fall back on Zarathustra. In a notebook of the period Nietzsche wrote, 'You speak falsely of occurrences and accidents! Nothing will ever occur to you than you yourself! And what you call accident (*Zufall*) – you yourself are what occurs to you (*euch zufällt* – literally 'befalls you') and falls on you.'[10] This sounds like Zarathustra (the 'you' is plural, suggesting his discourses) and it expresses both the universal prevalence of chance and its corollary, the idea that Nietzsche calls *ego fatum*. We are not to identify ourselves with one or other side of the collisions of forces that are called 'accidents'. Rather, such interactions are what constitutes our existence in the first place. The dwarf is expressing much the same idea, though in a one-sided and hostile fashion. The difference is that he takes the conflict to be simply destructive, whereas for Nietzsche (and Zarathustra) it is precisely the realization of creativity. As we shall see, however, the relation between the spirit of gravity's utterances and Zarathustra's teachings raises issues over which later writers have disagreed.

The Mountain Path

At the beginning of Zarathustra's narrative, he is climbing a steep and rough path. According to his friend Reinhart von Seydlitz, Nietzsche once reported a dream involving such a path: 'Nietzsche said, laughing, that in a dream he had been climbing up an endless mountain path; high up under the brow of the mountain, he had wanted to go into a cave, when from the dark depths a voice called to him, "Alpa, Alpa – who is carrying his ashes to the mountain?"'[11] More recent experiences may also lie behind the picture of the mountain path. Part Three of *Thus Spoke Zarathustra* was written in Nice, where one of Nietzsche's local walks involved climbing an old mule path which rises from the coastal road (the Roman Via Aurelia, now the Basse Corniche) to the hill village of Èze, four hundred metres higher.[12] Anyone who has undertaken the 'sentier Frédéric Nietzsche', as it is now named, will recognize Zarathustra's description, loose stones and all – but may, I think, be sceptical about Nietzsche's claim to have engaged in literary composition while occupied with this strenuous activity.[13]

Sitting on Zarathustra's shoulder is his 'arch-enemy' the spirit of gravity, half dwarf and half mole, engaged in pouring 'leaden thoughts' into his ear. The dwarf is malicious, but not subtle. He reminds Zarathustra that whatever goes up must come down, and his scornful whisper is echoed by a clatter of pebbles underfoot. The narrow path is not only difficult but dangerous, since one might easily slip and fall into the abyss. This dizziness caused by looking into the abyss, usually called a fear of height, is really a fear of depth. It is a danger for the spirit of gravity who, unable to overcome his fear, can only crouch when set down on the height, as Zarathustra points out, playing on the words *hocken* (crouch) and *hoch* (high). On closer inspection, the situation is more complicated. In another chapter, Zarathustra has already described the danger confronted by the climber: 'Not the height but the precipice is terrible. That precipice where the glance plunges *down* and the hand reaches *up*. There the heart becomes giddy confronted with its double will.'[14] According to this, what confuses the mind is a dissociation of glance and hand. The immediate threat for the climber is an inner one, a result of conflicting influences within oneself.

Just what are these influences? Can we be more specific about their nature? An earlier notebook draft, dating from 1875, contains a striking parallel to this passage. As he often does in later writings, Nietzsche compares the thinker with an explorer, and remarks:

> What is astonishing in careers of this sort is the way in which two hostile drives, which press in opposite directions, are constrained to proceed under a *single* yoke, so to speak. The drive which desires knowledge must again and again leave the inhabited lands behind and venture forth into the unknown; and the drive which desires life must again and again grope its way back to an approximately secure place on which it can stand. We are reminded of James Cook, who for three months had to feel his way across a chain of reefs with

his sounding-lead in his hand, and for whom dangers often increased to the point where he was glad to return for refuge, even to a location which he had shortly before considered to be most dangerous.[15]

The reference here is to Cook's voyage of exploration along the eastern coast of Australia in 1770. As Cook followed the shoreline northward his ship, the *Endeavour*, became trapped behind the Great Barrier Reef in an area of dangerous rocks and shoals. After three months, Cook managed to find his way out to the safety of the open sea. A few days later, having sailed westward to remain within sight of land, he was forced by strong currents on to the reef, and barely managed to pass through it again without being wrecked. This anecdote was found by Nietzsche in a biographical essay on James Cook by a contemporary admirer, G. C. Lichtenberg.[16] According to Lichtenberg, the main feature of Cook's character was an unshakeable determination, 'bordering on stubbornness', to carry through any undertaking.[17] The story of his return to the Great Barrier Reef, from which he had just escaped, is strong evidence for the claim. On that occasion, Cook did face the dilemma described by Nietzsche: between a drive for knowledge – that is, his determination to locate the northernmost point of the Australian mainland – and the drive to self-preservation.

Nietzsche describes those two opposed drives as 'constrained to proceed under a single yoke', like the white and black horses in Plato's *Phaedrus*, which draw the charioteer of the soul in opposite directions, upward or downward.[18] The tension is a source of danger, but it is also an opportunity for creative development, even a necessary condition for that end. The conflict of forces within the individual, or between individual and group, is identified by Nietzsche as the origin of the greatest achievements. He argues for this thesis in respect of ancient Greek culture in his 1872 essay 'Homer's Contest' and, more importantly, in *The Birth of Tragedy*, where Greek tragedy is explained as the product of an interaction between the opposed 'Apollonian' and 'Dionysian' drives. In the 'free spirit' period, Nietzsche sees the conflict in naturalistic terms, but still affirms its productive function. In *Daybreak*, for example, the 'so-called genius' is explained by a 'physiological contradiction' between two kinds of inner impulse, one disorderly and random, the other highly purposeful, existing side by side, bound together yet often in conflict, and both essential to the production of the work of art.[19]

This is not to say that every conflict of impulses is a creative or productive one. A one-sided contest will not lead to anything higher, and an even match may only be mutually destructive. Certain conditions are needed for a favourable outcome. Just as climbing requires the co-ordination of eye and hand, so the process of self-overcoming advances only by the co-ordinated action of the powerful forces within personality. Zarathustra goes on:

Alas, friends, can you guess what is my heart's double will? This, this is *my* precipice and my danger, that my glance plunges into the height and that my hand would grasp and hold on to the depth. My will clings to man; with

> fetters I bind myself to man because I am swept up towards the overman; for that way my other will wants to go And therefore I live blind among men as if I did not know them, that my hand might not wholly lose its faith in what is firm.[20]

Here the usual situation of the climber is inverted. Zarathustra's abyss is the clear and pure heaven over him, an 'abyss of light'. 'To throw myself into your height, that is *my* depth.'[21] Again, the gaze is an attraction and a temptation, perhaps like the 'action at a distance' that Nietzsche associates with the feminine element.[22]

The other danger suggested by the mountain path is a risk of falling, made greater by the spirit of gravity who sits on Zarathustra's shoulders, drawing his foot downward towards the abyss. The distinction between height and depth, peak and abyss, is not just a contrast, but a conflict which anyone who climbs the path must deal with. Life requires height, Zarathustra tells us. 'And because it requires height, it requires steps and contradiction between steps and climbers. Life wants to climb and to overcome itself climbing.'[23] The struggle against obstacles that stand in one's way is one conflict: 'Crushing the rock that made it slip, my foot forced itself upward', says Zarathustra. As well as this, there is a conflict *within* the climber, the 'double will' that represents a tension between forces acting in upward and downward directions. All this, it seems, is necessary for attaining height, because the steps over which we make our way are just the stages we have already achieved. The ascent is a process of self-overcoming.

Unlike the lanes that meet at the gateway 'Moment', the mountain pathway is not a way for everyone. Zarathustra observes that 'fewer and fewer climb with me on higher and higher mountains'.[24] Only those who are able to climb, and who have the determination and courage that climbing demands, may proceed along this way. There is something *übermenschlich* about these adventurers; they are at least 'higher men', if not higher than men.[25] In *Schopenhauer as Educator* Nietzsche draws a related picture in envisaging a choice made by the contemporary individual arriving at a crossroad, like the Greek story of the young Hercules choosing between the ways of pleasure and virtue.[26] Anyone choosing the broad way of social conformity will join an 'enormous crowd' of contemporaries, offering mutual reinforcement as they proceed at the word of command. In contrast is one less travelled by, the way of the individual thinker:

> The other way will bring him together with fewer fellow travellers, it is more difficult, more tortuous, steeper; those who walk on the first way scorn him because he steps there more wearily and is more often in danger, and they try to entice him over to themselves. If the two paths happen to cross he is mistreated, thrown aside or isolated by being cautiously walked around.[27]

As these passages show, the mountain path image recurs in Nietzsche's writing. It is always identified as a difficult and laborious way, travelled by the relatively few who have the courage to undergo the dangerous conflict of forces there. The path in *Thus Spoke Zarathustra* retains these connotations but is presented in juxtaposition to other ways, with which it stands in contrast. The lanes in 'On the Vision and Riddle' are not unlike the common way described in *Schopenhauer as Educator*, but they also have a temporal meaning, providing a model of time that interprets becoming in terms of past, present and future. So, can it be that the pathway also has a symbolic function of representing temporality?

The Temporality of the Mountain Path

As Zarathustra emphasizes, climbing involves effort and fatigue. This could simply be attributed to the struggle to force oneself upwards, and yet, as any climber knows, a descent can be as difficult and dangerous as an ascent. What the two have in common is that one must continually be starting and stopping. The need to begin again at each step is a repeated demand on the climber which, even with no added burden, is enough to bring about a sense of oppression and boredom (the image of Sisyphus is not far away here). Yet this is inescapable, given that 'climbing' implies a succession of steps, taken one at a time. Not surprisingly, the idea is also expressed in images of stairs or ladders.[28] Movement along this way is a sequence of discrete stages, not a continuous process. Even the spirit of gravity, who is not climbing but riding on a climber's shoulder, conforms to the pattern by uttering his words 'syllable by syllable'. As Nietzsche explains in his Basel lectures, the syllable is the elementary unit of speech (corresponding to the single step, mimetic gesture or musical tone) that determines the time-unit of rhythmic form.[29]

A model of time is indicated here: the path stands for a temporality constituted by finite intervals. Furthermore, climbing suggests both starting and stopping points, such as the 'peak' at which the ascent reaches its limit (although there are 'abysses' above and below as well). Applied to time, this implies a determinate, finite duration, with a further infinitude on either side. Time is taken as something measurable and countable in units, such as those indicated on a clock – whose hands, as Nietzsche often notes, usually advance in steps, pausing before moving forward again.[30] There is a strong link with his earlier reflections on rhythm and meter, which presupposed minimal units of time (*chronoi protoi*) as the basis for patterns of long and short intervals within a given composition. The rise and fall of the foot was the prototype for these temporal forms, as it is here.

In Plato's terminology, the mountain path is about the *parts* of time, and the gateway and lanes about the *forms* of time. Now we need to recall that Nietzsche's argument against the possibility of a state of rest, as well as his argument for an eternal recurrence, rely on the assumption that whatever is possible must occur within an infinite time, a premise whose most plausible support is the Archimedean principle concerning the addition of finite quantities. If only a finite number of states of affairs are possible, then the total time of their occurrence must be a finite time. This in turn assumes that any given state lasts for a finite time, with the sole exception of the thinkable but impossible state of rest. The model of gateway and lanes gave no basis for finitude, but the pathway's temporality is just where this concept is to be found.

What are these finite intervals? Some writers of Nietzsche's time postulated a natural measure of time for every living species, and hence for every sentient individual. The thesis was presented with dramatic flair by the biologist Karl Ernst von Baer in a text cited by Nietzsche in his Basel lectures. Baer's address on 'the correct approach to living nature' was delivered in October 1860 – that is, just a year after the appearance of Charles Darwin's *The Origin of Species.* Never mentioned by name, Darwin is an implied presence throughout, for Baer is engaged in a rearguard action against the perceived advance of materialism in biological science. It is hard to believe that Nietzsche did not identify this purpose, with which he had some sympathy, although he would not have endorsed Baer's rival model of ideal types. In a preface, Baer states his general thesis: the state of a living thing at any moment can be explained only by its total life-process, and not by material causes.[31] This is why the specificity of time-awareness is explored at length in his discussion. A structuring of time is central to the life-process, in both its physical and mental expressions.

Nature, Baer says, is a process of constant change, and living things are constantly developing and interacting. 'A stability (*Beharren*) does not exist anywhere in nature, and least of all in living things. It lies only in the small measure that we apply when we believe that we observe a stability in living nature.'[32] Baer notes that we see the world on a human scale: for example, traditional units of length refer in some way to our bodies. He then goes from space to time, asserting that different living beings experience the flow of time in quite different ways, according to their constitutions. Every organism has a life-process which proceeds at a certain pace of its own. Baer uses the pulse rate as a sign of this pace, although he mentions other indications such as the typical natural lifespan. He then illustrates the relative character of our experience of time by posing a sequence of thought experiments, involving different rates of the life-process.

If human life were speeded up, we would see natural processes as slower than they now seem to us (assuming the same senses in operation). In fact, this is not an imaginary situation: it is the experience of those species whose lives are counted in days or hours. Baer takes his audience through several stages of reduction in scale, coming down to a conscious lifespan of only 40 minutes. In that case, he says, one would have no conception of day and night as alternating: the onset of evening would seem the end of the world. Grass, flowers and trees would seem to be changeless realities. The sounds that we now hear would be imperceptible, yet other sounds would be heard for the first time. Baer concludes by emphasizing these qualitative contrasts: 'If our innate measure of time were different, external nature would necessarily appear quite different, not just shorter or longer in its processes and narrower or wider in its effects, but different throughout.'[33] Conversely, if our life-process were slowed down, we would see events as occurring with constant change, and lose any sense of permanence. Again, Baer proceeds in several steps, each time increasing the imagined lifespan a thousandfold. With a long measure, the cycle of day and night would vanish. The sun would be a glowing band across the sky, and the seasons seem mere fluctuations. As before, these observations appeal only to our familiar human senses: if one adds the possibility of other modes of perception, which we know exist in other species, the differences would be even more striking.

Is there any way of saying which of all these scales is the correct one? Baer writes, 'Which view is more correct, or comes closer to the truth? Without doubt the one which proceeds from the greater measure. Nature works with unbounded time in unbounded space. The measure for its activity can never be too big, but is always too small.'[34] Thus, the truest timescale is the infinitely great one, according to which even the longest duration known to us is only a fleeting moment. The Heraclitean interpretation of the world as a continual flow now appears justified by scientific thought. Stability is an illusion: what is truly permanent is just change, and in particular the kind of change that we call 'development'– a word that, for Baer, signals the governing role of laws of nature and of ideal biological types.

How impressed Nietzsche was by this presentation can be seen in his lectures on pre-Platonic philosophers, which borrow from Baer in explicating the Heraclitean doctrine of becoming. Nietzsche explains, 'The *inner* life of different living species (including human beings) proceeds in the same astronomical time with specifically different speed: and the different subjective measure of time is governed by this.'[35] The theme is illustrated by Baer's two thought experiments.

> If we could perceive still faster, the illusion of permanence would be even stronger; if one imagines the infinitely fast, but still a human perception, then all motion ceases, everything would be eternally fixed. On the other hand, if one imagines human perception infinitely increased in the strength and power of the organs, then conversely nothing enduring would appear, even into the infinitely smallest part of time, but only a becoming. For the infinitely fastest perception all becoming ceases, if human perception is still intended. If it were infinitely strong and pressed into every depth, then every form would cease: only for a certain degree of perception are there forms.[36]

That is, if our pulse rate were speeded up, our subjective experiences would be greatly increased in number. Taking this tendency further, we would see everything as permanent, and have to infer change as we now infer the motion of the heavenly bodies. For an infinitely fast perception, movement would cease altogether. In contrast, if our lives were extended greatly and our perceptions slowed down accordingly (Nietzsche describes this as a 'strengthening' of our capacities) permanence would be taken as mere illusion, and even the determinate forms of things would disappear. The actual human experience of the world corresponds to a certain degree of perception, somewhere between these two extremes.

As we noted, Baer gave a biological identification of this measure with the rate of the heartbeat which corresponds to the tempo of the animal's life-process. Every species has a natural rhythm for its metabolism, just as it has a natural configuration of its organs. Baer's observations about the human species add more detail to the significance of this specific temporality. Human beings are like animals physically but not mentally, he asserts: they have a spiritual life that the materialistic researcher fails to acknowledge. The relation between the physical and the mental is like that between a written score and the music that it symbolizes. A savage who sees the score, Baer explains, will not know what paper is, or else may recognize paper but attach no meaning to the lines and other marks on it. A civilized person knows that these stand for music, but may be unable to read them. Only an educated musician will recognize the music as a composition of Mozart or Beethoven.[37] The materialist investigator is like the savage, in that he sees the physical signs but has no awareness of the corresponding spiritual reality. Yet that is, in fact, what comes first and determines the form of its material existence; for the thought chooses its spoken words, and the melody its heard tones.[38] More generally, Baer argues that the life-process of an organic being is like the performance of a musical composition. As with melody, there is a harmony, present in what Baer calls the 'type' – that is, the pattern determining the spatial relations between parts of the organism. Similarly, the life-process has its own tempo and rhythm, and it develops through successive stages according to its own laws, not those of matter.[39] In his own scientific research in embryology, Baer identified several 'laws' governing such development across species. The structuring of time, and the regularity of recurring patterns, are central themes here.

Baer is not saying that the pulse rate of the animal determines the tempo of its life-process, but rather that the ideal type is what explains the observed biological phenomena, with the pulse rate as the most characteristic instance. That is the main point of the musical analogy. It is music as conceived by the composer, not as heard by an audience, that he has in mind, in contrast with Nietzsche's uses of the musical score image.[40] Faced with this thesis, clearly aligned with philosophical idealism, the appeal of Baer's argument for Nietzsche would run up against his strong naturalistic tendency. In the end, he regards the interplay of forces within the organism as the determining factor for all facts of consciousness, including temporal awareness. This physiology may escape our observation, but that does not make it something ideal.

Is There a Measure of Time?

Is there a psychological as well as a physiological unit of time for the living being? Gustav Teichmüller, another contemporary writer admired by Nietzsche, argued that the natural unit of our temporal perception is the so-called present, which has a certain extension although we naturally think of it as indivisible. As techniques of measurement improve with scientific discoveries (in electricity and magnetism, for instance) we are able to subdivide this time further and further. 'Then we find that the time that we distinguish from past and future as the present moment has a certain quantity or duration, and is not at all an indivisible point.'[41] But even if a subjective 'Now' can be divided into an indefinite number of shorter timespans, what we encounter in direct experience is the 'specious present', as William James called it,[42] and that determines the standard by which other processes are interpreted as taking a long or short time. Similarly, J. F. Herbart suggests a psychological unit of time roughly comparable with our smallest conventional units, that is, minutes and seconds.[43] For other beings, this natural unit of time could be quite different.

However, for practical purposes we need to systematize our measurement of time. Nietzsche agrees that this operation is bound up with recurring processes, and for that reason relies on assumptions in a way that measurement of space does not. We can move a rigid ruler about and place it against an object we want to measure, but this cannot be done with a process taken as the measure of time. For this purpose we appeal to processes that occur at a constant rate – that is, which produce the same amount of change (for example, motion through the same distance) in the same time – in order to say that earlier and later intervals of time are similar in length. Hence, the significance of naturally occurring rhythms such as heartbeats or astronomical cycles. Living beings like ourselves undergo a number of recurring cycles – the day, the month, the year – which are reflected in our metabolism and in our forms of life. We also have clocks, but Nietzsche sees these as representing social convention (or 'law') compared with natural signifiers of time such as stars, sun and shadows (which all move continuously, not in distinct steps).[44]

In a note on eternal recurrence, Nietzsche invites the reader to reflect on the effects of the annual cycle of seasons, or periodic illnesses, or waking and sleeping.[45] These cycles are determined by the alternating dominance of one or other of two opposing forces, as he explains in his lectures, adapting a passage in Plutarch: 'The occurrence of life and death, or waking and sleeping, is just the visible preponderance which one force has gained over its opposite and immediately begins to concede again to the latter.'[46] Nietzsche does not say what he expects to be learned from these experiences of recurrence, although an allusion in *Ecce Homo* to his own history of 'the periodicity of a kind of *décadence*'[47] suggests a reference to patterns of fluctuating vitality, linked with environmental conditions such as climate and weather, yet not showing any marked regularity, as one sees in his own history of unexpected recoveries and periods of well-being such as the 'Sanctus Januarius' of 1882.

Do any of these experiences provide an objective measure of time? Rather, we see everything measured by everything else. Yet, as we noted earlier, it is hard not to suppose that there must be a real time, even if it is unknown to us. It is with this in mind that Nietzsche observes in an 1881 note:

> To the actual course of things must also correspond an *actual* time, quite apart from the *feelings* of longer and shorter intervals that we knowing beings have. Actual time is probably unspeakably *slower* than we human beings feel time: we perceive *so little*, although even for us a day appears very long compared with the same day in the feeling of an insect. But our blood circulation could in fact have the duration of a revolution of the earth and sun. – Hence we probably feel ourselves as much bigger than we are, and so overrate our importance, in that we introduce too great [*sic* – apparently a slip on Nietzsche's part] a measure into space. It is possible that everything is much smaller. Thus the actual world *smaller*, but *moving much slower*, but infinitely *richer in processes* than we suppose.[48]

He starts here by postulating that there is an 'actual time'. In other words, time is not just a subjective feeling, but an aspect of 'the actual course of things'. In that case, what is the most accurate representation of things in time? Nietzsche's answer seems to be: the one that corresponds to the greatest power of perception. Consider, in contrast, living things that lack sentience altogether. Without sensations of pleasure or pain, they have no motivation to acknowledge change. Hence, if a plant could think it would be an Eleatic philosopher: "To the plants all things are usually at rest, eternal, every thing identical with itself . . . It may even be that the original belief of everything organic was from the very beginning that all the rest of the world is one and unmoving."[49] At the other end of the scale one would have to imagine a being that perceives everything, although this must be impossible, since it would imply an infinitude of content. Human beings are somewhere in the middle.

Nietzsche's conclusion is that the best measure of time is a much smaller one than ours, corresponding to the countless brief processes that we usually overlook. In that case, the processes of our everyday experience take much longer than we suppose – that is, they are 'unspeakably slower'. A day seems long to us, but even longer to an insect for whom it may be an entire lifetime. Nietzsche speculates that it may really be longer still. On the other hand, he seems to be saying that the best measure of space is a much bigger one than we commonly use, so that we ourselves (that is, our bodies) are really much smaller than we assume. This contrast between space and time recalls the time-atom theory's assertion of their interdependence: we *need* space in order to account for the things that we take to be in concurrent existence, although that belief is due to our limited awareness of succession.

Several mistakes arise from these perspectival impressions. First, we think that many events occur suddenly. 'There are an infinite number of processes that elude us in this second of suddenness.'[50] Secondly, we imagine that two events are simultaneous when they are not. The time that separates them is negligible only by a measure which is too great for the actual content of becoming. We perceive very little of what happens and so take processes to be sudden when, in fact, they are quite slow. This is what Nietzsche means when he suggests that our 'blood circulation' (meaning our heart rate) could be as long as an astronomical cycle – at any rate, far longer than the rhythm of more basic processes that we fail to perceive. On this view, everything we perceive is much slower or gradual than we suppose. Sometimes we gain a sense of this truth, Nietzsche says, when time seems to slow down: for example, under the influence of hashish, or in moments of mortal danger.[51] In these states we have thousands of thoughts and experiences within a single second, as measured by a watch. In effect, we are like the short-lived beings of Baer's thought-experiment. By our ordinary standards, each distinct event that they experience takes next to no time at all. This is presumably what Nietzsche means when he writes, 'When one looks into the realm of the infinitely small, every development is always an *infinitely fast* one.'[52] By the same token, what we ordinarily perceive as single events turn out to be much more extended – that is, to be slower.

Here we are imagining replacing our usual measure of time with a much smaller one. But according to Baer, the longest scale must be taken as the most reliable one, for both space and time. Compared to this, everything that we experience is very fleeting. On that premise, everything is much faster than we suppose. Baer concludes that the Heraclitean view of things must be regarded as the correct one, namely, that everything is eternally in flux.[53] Which view is right? It depends on whether one is considering the objects of everyday experience in a new perspective or, as Nietzsche wants, setting them aside in favour of a completely different range of objects, 'the realm of the infinitely small'. If an hour seems to go very slowly, this is because the events that one is now aware of are 'infinitely fast' from the usual perspective. It is likely that Nietzsche's real preference is for this multiplicity of perspectives, rather than taking one or another as having a privileged status. In that case, adopting one or other extreme perspective, infinite or infinitesimal, is a way of challenging the seeming givenness of our human timescale. It is not that some other temporal perspective is the 'true' one, but rather that every perspective is to be assessed as a function of the form of life that it both expresses and promotes. This important point will need to be borne in mind in what follows.

The Dynamics of the Mountain Path

So far we have seen that a temporality of finite periods is symbolized by the mountain path, and that motion on this way (that is, climbing) is characterized by a perpetual tension between opposing forces. A task that still remains is to explain the connection between these. Nietzsche's interest in rhythmic processes, informed by theorists as diverse as Aristoxenus and K. E. von Baer, is a good source of clues. The sense of time that sets up a measure for assigning magnitudes to intervals is grounded in the life-process of the conscious being. This in turn involves the interplay of forces within the organism or between it and its surrounding environment. What is there about climbing that could give rise to definite periods of time? It must involve the opposing forces at work there, and so their interaction needs to be examined more closely.

Each stage of climbing involves a conflict or 'contradiction' between the climber and the step traversed. In an earlier chapter, Zarathustra encounters a young man who expresses himself in these words: 'I no longer trust myself since I aspire to the height, and nobody trusts me any more; how did this happen? I change too fast: my today refutes (*widerlegt*) my yesterday. I often skip steps when I climb: no step forgives me that.'[54] A similar expression appears in a notebook of the same period: 'You contradict today what you taught yesterday – But that is why yesterday is not today, said Zarathustra.'[55] Nietzsche did not use the idea in this wording, perhaps because it is too obviously modelled on Emerson's advice in 'Self-Reliance' about inconsistency: 'Speak what you think in hard words; and tomorrow speak what tomorrow thinks in hard words again, though it contradict everything you said today.'[56] The immediate point is linked with his repudiation of 'convictions' as barriers to self-development, but this is one aspect of the 'metamorphoses of the spirit' through distinct stages. What we speak or teach expresses what we are, revealing whether this is a stable 'character'[57] or something open to continual change and transformation.

Other passages in Nietzsche repeat the theme of contradiction and time. 'More and more it seems to me that the philosopher, being *necessarily* a man of tomorrow and the day after tomorrow, has always found himself, and *had* to find himself, in contradiction to his today.'[58] Similarly, the 'new philosophers' of *The Gay Science* are described as 'stretched out (*hineingespannt*) in the contradiction between today and tomorrow'.[59] With one foot on either side, so to speak, the untimely thinker experiences these contradictions as an inner tension: hence, the youth's distress. There are important contrasts between this dynamic relation between parts of time and the conflict between the forms of time, or 'lanes'. Whereas past and future confront each other within the gateway 'Moment' as wholes, here the disagreement is between particular elements of time: for instance, between yesterday and today, or between today and tomorrow. The conflict is between earlier and later, not between past and future. Once again we see the importance of making this distinction. The comment 'that is why yesterday is not today' hints that a given day is succeeded by the next because of a conflict, exemplified in the contradiction between one 'teaching' and another.

What is the relation between contradiction and succession in time? One answer is given in Kant's statement that 'Only in time can two contradictorily opposed predicates meet in one and the same object, namely, *one after the other.*'[60] Schopenhauer repeats the point: 'Therefore time can also be defined as the possibility of opposite conditions in one and the same thing.'[61] For instance, the propositions 'Socrates is sitting' and 'Socrates is walking' are consistent with each other, if taken as true at different times – for example, on different days. This requires us not only to recognize a temporal difference, but also to assume the continuing existence of a thing that is the subject of change, here an individual person. We compare our perception of this object's present state with our memory of its previous state, which was different. Because these descriptions are inconsistent, one is assigned to the past in order to resolve the contradiction. This solution is supported by a metaphysical commitment to enduring things or, in Aristotelian language, to the concept of *substance.* According to Aristotle, 'It seems most distinctive of substance that what is numerically one and the same is able to receive contraries.'[62] It does this by changing over time: 'that which was hot becomes cold, for it has entered into a different state'.

This form of temporality appears as the consequence of an interplay of forces, just as the climber's movement from one step to the next is due to a conflict between opposing forces. The cyclical rhythm of the life-process which gives rise to a natural measure of time for each species is due to a characteristic interplay of physiological forces. Baer's example of the animal heartbeat, consisting in regularly alternating phases of muscular tension and relaxation, is an apt example. This brings us back to Nietzsche's interpretation of Greek rhythm, which continued to influence his thinking for many years. Rhythm is not just a succession at regular intervals, producing a patterned division of time. Rather, it is an expression of a process that has a characteristic period of occurrence. The raising and lowering of the foot involves an interaction between forces, as does the sway of a pendulum and its human counterpart, the motion of the conductor's hand.

What is it that determines the finite interval that constitute the measure of time for the living beings – or, for that matter, any other finite interval? On our usual understanding, the only relation between earlier and later states of affairs is a causal connection. In that case, the amount of time between them is just a given fact: a certain cause gives rise to an effect after a certain interval. There are problems here, as we noted in Chapter 2. Why should there be any delay at all, if the cause is a sufficient condition of the effect? What is more, this model does not fit the phenomena of rhythm well, especially in living things. As Baer might have objected, one heartbeat is not the *cause* of the next one. Rather, they arise out of the underlying life-process, are subject to its variations and intermittences, and it is just when that support fails that they cease to occur. This is Nietzsche's view as well, only without the idealistic postulate of a timeless archetype. Instead he insists that the body is the location of the governing power. As Zarathustra puts it, 'There is more reason in your body than in your best wisdom.'[63]

For Nietzsche, then, all temporal finitude is a function of a conflict or struggle of forces. As we saw earlier, this is an important link between becoming and time. Every conflict takes time to occur, whereas the concept of causality has no such guarantee built into it. Nietzsche takes the same idea further when he writes, 'All occurrence, all motion, all becoming as an establishing of relations of degree and force (*ein Feststellen von Grad- und Kraftverhältnissen*), as a *struggle* . . .'[64] Here he is specifying the nature of the conflict, or rather, of its outcome. It has to do with a disparity in power which provokes a contest for domination and the establishing of an 'order of rank'. What is meant by *Feststellen* is not immediately evident, but earlier in the same note the 'will to truth' is described as 'a making firm, a making true and durable (*ein Fest-stellen, ein Wahr-Dauerhaft-Machen*)'. These stabilities may be temporary, but they allow us to identify relatively lasting situations. Hence the rests, and the assignment of becoming to the transitions from one state of rest to another.

The account presented in this chapter may be summarized along the following lines. What we encounter in understanding time as consisting of finite intervals is a particular interpretation of becoming. Time arises when the conflict of force that is essential to becoming is conceptualized according to a temporal schematism. In the imagery used by Nietzsche, the condition of conflict is a polarity of force, here defined as upward and downward. Aspiring to the height and drawn into the abyss: this is Zarathustra's struggle, an inner tension between opposing impulses giving rise to disagreements between eye and foot, or glance and hand. The conflict is resolved with every step, only to break out yet again. Temporal form and content, a duality not found in absolute becoming, are separated by this schematism, so that time comes to be understood as something in its own right, along with parts of time such as days or years.

So far, these collisions of force have been manifested in experiences of resistance, disturbance and turmoil. There is another way to represent them, at the level of conceptual thinking and discourse. The relation between the distinct parts of time (symbolized as the steps that constitute the way for climbers) has been defined as a *contradiction*. There must be as many of these contradictions as there are distinct periods of time, since each expresses the transition from one state to another as a contradiction between two descriptions of the same thing. But this is not the only kind of contradiction involved in models of time. Another is the contradiction between the forms of time, past and future. The next chapter will follow the narrative of 'On the Vision and Riddle' from its first phase, set on the mountain pathway, and the predicament of Zarathustra as he is beset by conflicting pressures, to a new situation that introduces a completely different model of temporality, one that presents its own issues.

Chapter 5

Gateway and Lanes

We are still on the mountain path in the gloom of twilight. After a while the dwarf runs out of ammunition and falls silent as Zarathustra continues climbing, although weary and oppressed by evil dreams. Summoning his courage (a useful quality, as he digresses to remind his audience) he halts and confronts his companion, warning him that 'you do not know my abysmal thought. *That* you could not bear!' In response, the dwarf jumps off his shoulder and crouches on a stone, curious to find out this thought. Once relieved of his distracting burden, Zarathustra is able to assert himself and take control of the situation. The scene is set for another confrontation between two figures, as with Heraclitus and Parmenides on the way leading to the 'abyss of all things'.

Zarathustra's words have signalled a new situation. There is a gateway where he has stopped.

> 'Behold this gateway, dwarf!' I continued. 'It has two faces. Two ways come together here: no one has yet followed either to its end. This long lane stretches back for an eternity. And that long lane out there, that is another eternity. They contradict each other, these ways; they offend each other face to face; and it is here at this gateway that they come together. The name of the gateway is inscribed above: 'Moment'. But whoever would follow one of them, on and on, farther and farther do you believe, dwarf, that these ways contradict each other eternally?'

Zarathustra's question involves as an essential element the task that is typical of the riddle in its purest form, namely, the deciphering of a metaphor. As Aristotle points out, good riddles and good metaphors go together.[1] The question asked by Zarathustra can be answered only if its symbols are interpreted adequately. With this in mind, we need to look closely at the situation that it depicts. The image of gateway and lanes seems a straightforward metaphor, taken by itself. The gateway's literal meaning is conveyed in Zarathustra's statement that its name is 'Moment'. It has two faces, one looking forward and the other backward. Accordingly, the two lanes on either side must stand for the time before and after the present moment – in other words, for past time and future time.

One issue here is the infinitude of the two lanes, that is, of past and future time. On this point Nietzsche's position is clear: time must be understood as infinite in both directions. But that is mentioned only as the presupposition for a question which is less easy to understand, involving a confrontation or 'contradiction' between the two lanes, occurring within the gateway 'Moment'. The question raised is whether the lanes are in conflict 'eternally' or only in this gateway. Much of this chapter will be occupied with deciphering these metaphors, in the hope of finding a solution to the riddle posed by Zarathustra. The gateway and lanes must be considered together, since they are interdependent. Without the gateway, there would be a single way, not two distinct lanes. Without the two lanes, there would be no gateway, since it is just the transition from one lane to another. Still, we need to start somewhere, and so I begin by considering the gateway.

The Structure of the Gateway

As the word *Thorweg* indicates, this too is a way. Plato observes of the 'One' that 'of course, as it travels from past to future, it will never overstep the present'.[2] Anyone who goes from one lane to the other must pass through this gateway. It does not form any kind of barrier, therefore, but rather provides access. Further, the gateway has properties of its own. It has a lintel above on which the word 'Moment' is written. It also has a threshold (*Schwelle*) below, according to an earlier text in which Nietzsche states: 'One who cannot set himself down on the threshold of the moment, forgetting all that is past, who cannot stand on one point like a goddess of victory, without giddiness or fear, will never know what happiness is . . .'[3] The gateway 'Moment' is thus like the one described in the philosophical poem of Parmenides: 'There are the gates of the ways of Night and Day, fitted above with a lintel and below with a threshold of stone.'[4] A difference is that Zarathustra's gateway has no hinged gates that can be closed and fastened to bar the passage of an unauthorized traveller. These lanes and the gateway between are common to whatever participates in time by being future, present and past, and not reserved for some only. The openness of this way is, in fact, one of its essential features.

The gateway is a sort of *frame* for the way within it. This structure is not itself the present moment – rather, it encloses the moment. Such a frame allows a direct confrontation of past and future in the moment: they come together and meet face to face, not by proxy. If past, present and future were parts of time, there would be no direct encounter between past and future. Instead, they would be separated by the present. But the 'moment' does not *separate* past and future, because they come together within it. This point is made clearly by Aristotle, who observes that the 'now' is something that divides past from future time, but also links them, rather than keeping them apart.[5]

But how are we to understand a 'coming together' of past and future? Perhaps in the following way. We commonly use the word 'present' to refer to some part of time, such as the present year, which stands between past years and future years. However, there must be a present moment, according to an argument of Aristotle which starts from the observation that the present year can be divided between the present day and other days that are past and future, like the days of other years. In fact, as soon as parts of the year are recognized, one must be taken as the 'present'. Similarly, in looking closely at the present day, we find the present hour, as well as past and future hours. If time is indefinitely divisible, then this line of thought can be continued without end.[6] Ultimately a temporal point is reached, and this is what is denoted by the word 'now' when it is used in the strictest sense, the one from which, Aristotle concludes, other uses – for the present hour or day, for instance – derive their justification.[7]

This may look like convergence on an unextended moment, but it cannot be what Nietzsche means by the 'coming together' of past and future in the gateway 'Moment'. The only process described is that of our own thinking about time, and that does not contain anything corresponding to Nietzsche's images of conflict and impact. Still, this analysis of the concept 'present' identifies the *place* of struggle between past and future, and so is relevant to the dynamic approach set out later in this chapter. It confirms that the gateway (which is not the frame that encloses it, although the same word is used for both) is unextended, in contrast with the lanes.

The gateway has further properties of its own. Zarathustra says that it has two faces, suggesting that these are dissimilar in character. The two faces of the gateway determine what the lanes are that meet there: the one by the forward-looking face is the way of the future, while the one by the backward-looking face is the way of the past. Without the presence of the gateway there would not be *two* lanes at all. The concepts 'past' and 'future' make sense only in relation to some present moment which, by having two aspects, creates that distinction. But how does it do this? One clue is the name of the gateway. The German word translated as 'moment' is *Augenblick.* This literally means a glance of the eye, a familiar event that occurs in next to no time.[8] More broadly, it suggests a point of view. In this sense, the moment is a perspective on a temporality consisting of a past and future, because it looks forward along one of the ways and backward along the other. Thus, the gateway is not simply the place at which the lanes happen to meet and 'contradict' each other. It is what *makes* them two lanes and sets them in opposition to each other. Without this there would be only one way, having no division within itself.

The claim that the two lanes 'come together' at the gateway is not just a kinematic abstraction: their coming together is depicted as a confrontation or collision. How are we to understand the idea that the two ways 'contradict each other' and 'offend each other face to face'? These are unlikely descriptions of lanes, but let us consider each in turn. When two statements contradict each other, they cannot both be true. What does that have to do with past and future? And how does it differ from the contradictions discussed in Chapter 4? The observation that future states of affairs differ from past ones is hardly sufficient to constitute a contradiction in this sense. In the first place, it is just temporal differences that allow different descriptions of the same things without contradiction. The concept of change requires the total state of the world to be different at different times: without that, time itself would become meaningless. In any case, later states of affairs differ from earlier ones within the past, taken by itself, and within the future, taken by itself. So a contradiction identified with the present moment must have some basis in the relation of past and future as totalities, rather than being due only to differences over periods of time.

We should remember that the word 'contradict' refers not only to a logical relation between propositions, but also to a dialogical relation between speakers. People who disagree contradict each other, and given Nietzsche's metaphorical mode of thought, he surely has this in mind. What occurs in the gateway is some *disagreement* between the two lanes; and so one part of Zarathustra's challenge is to discover its nature. What do the past and future 'say', and how do their statements contradict each other? No clues are given, and we must venture further afield to answer this question.

In his Basel lectures on the pre-Platonic philosophers, Nietzsche uses the expressions 'It was' and 'It will be' to indicate participation in time and becoming. His source is evidently Plato's statement that 'It was' and 'It will be' are the forms (*eide*) of time, whereas days, months and years are parts of time.[9] Accordingly, Nietzsche emphasizes the timelessness of the Eleatic conception of being by saying that according to this doctrine, 'what truly exists, exists in an eternal present, and it cannot be said of it, "It was", or "It will be". The concept of time has nothing to do with it.'[10] Similarly, when he argues that atomism eliminates genuine coming-to-be and passing away in favour of rearrangements of elements which are themselves permanent, he asserts that the Eleatic conception of being applies as well to these particles, so that 'it cannot be said of them, "It was", or "It will be"'.[11] With this in mind, I suggest that the opposing statements made by the two lanes of past and future are 'It was' and 'It will be'.

Further support can be found in considering uses of the expression 'It was' found in Nietzsche's published work up to and including *Thus Spoke Zarathustra.* In the second 'untimely meditation' he says that the child's discovery of the past comes with learning the meaning of the phrase 'It was'. The essay begins with a contrast between human beings and other animals, who are not subject to the cares that burden us, because their lives are contained wholly within the present moment. Their happiness depends on their inability to remember the past and foresee the future: because the past has no reality for them, it cannot be an object of regret, any more than the future can be an object of hope or fear. For these creatures, life is a continual flow of moments that arise out of nothingness and immediately pass back into nothingness. For human beings, time is very different. Not only are we aware of what has gone before, but we experience it as a weight that presses us down more and more. For this reason, we look with envy on the animals and on the small child who, like them, lives within the present moment, in a space protected by the 'fences of past and future'.

> Yet its play must be disturbed; all too soon it will be called out of its state of forgetfulness. Then it will learn to understand the phrase 'It was': that password with which conflict, suffering and weariness come upon man so as to remind him what his existence fundamentally is – an imperfect tense that can never be completed.[12]

It is through this event that we bring upon ourselves the 'great and ever greater burden of what is past'.[13] Nietzsche emphasizes the role of language: it is an understanding of 'It was' that marks the emergence of the historical sense, with all its consequences. He draws attention to the grammatical form of 'It was'. This is not just a past tense but an *imperfect* tense, and so it reminds us that human existence 'can never be completed'. Strictly, the point has to do with *aspect* rather than tense. The 'It was' is not something over and done with, which might be kept at a safe distance from the present. As Nietzsche puts it in a later notebook, the past has not passed away: *die Vergangenheit ist nicht vergangen.*[14] It presses through into the present and lays claim to it. Zarathustra's description of the gateway 'Moment' is a dramatic illustration.

The expression 'It was' plays an important part in *Thus Spoke Zarathustra*. There too it is linked with ideas of force and struggle. In the chapter 'On Redemption' Zarathustra contrasts past and future in terms of their relation to the will to create and transform, which makes possible all human achievement. The future is open to the will's action, whereas the past lies outside the scope of even the strongest will. Zarathustra says:

> Willing liberates; but what is it that puts even the liberator himself in fetters? 'It was' – that is the name of the will's gnashing of teeth and most secret melancholy. Powerless against what has been done, he is an angry spectator of all that is past. The will cannot will backwards; and that he cannot break time and time's covetousness, that is the will's loneliest melancholy.[15]

Whereas the future offers the will opportunities to satisfy its creative impulse by transforming its world, the past confronts the will as a power greater than itself. Accidental and fragmentary, it is 'the stone which he cannot move'. The phrase 'It was' is used to express the closed being of the past, placed in contrast with the open becoming of the future, which by analogy would be expressed as 'It will be'.

I have suggested making sense of the claim that the lanes of past and future contradict each other by supposing that they *say* 'It was' and 'It will be' respectively. Are there other possibilities? Going back to the work's opening chapter,

'On the Three Metamorphoses', one alternative might be that what the paths say is 'Thou shalt' and 'I will'. These utterances do contradict each other, and one is a hostile retort to the other, but it is hard to see any correlation with the two lanes. Alternatively, one might speculate that Zarathustra's words to the dwarf, 'I! or you!' correspond to the conflict within the gateway. That would mean identifying one lane with Zarathustra and the other with the spirit of gravity. There is no reason to do this and, in fact, the dispute between them is individualized in ways that do not match the conflict between the lanes. All in all, then, my interpretation appears to be the best way of grasping the contradiction between the lanes of past and future.

Yet 'It was' and 'It will be' contradict each other only if the 'it' is the same in each case. So, what is it that past and future are in disagreement *about*? It must be the present moment, since this is where the two lanes come together. On that assumption, their dispute is over whether the moment *belongs to* past time or to future time. That brings us to the other metaphor in this passage. In Walter Kaufmann's translation, the two ways are said to 'offend each other face to face'. This conveys disagreement, but not the image of head-on collision expressed in Nietzsche's words: *sie stossen sich gerade vor dem Kopf.* The two antagonists are pushing against each other. Each is trying to achieve something at the other's expense, that is, to gain ground by forcing the other back. They are contesting possession – but of what? Again, of the present moment. (Readers familiar with rugby football may want to imagine a scrum, where the forward players of opposing sides contend for possession of a ball thrown between them by the referee.) This reading is consistent with taking the 'contradiction' as a dispute over ownership of the moment: in fact, the two expressions reinforce each another, as Nietzsche surely intends.

We could, however, take these metaphors to support a stronger suggestion: that the claims made by past and future extend to *everything* that is in time. Although the moment is the immediate object of contention, the real demands go beyond it. The claim made by past time over time as a whole can be identified with the doctrine announced by a pessimistic 'soothsayer' in a previous chapter, that 'All is empty, all is the same, all has been (*Alles war*)!'[16] In asserting of everything that 'It was', this statement assimilates all time to the past. That implies that what is to come is no less accidental and fragmentary than what has been: the expressions of emptiness and indifference signal this, rather than the fixed and concluded character of the past. On the other hand, any attempt by the future to appropriate the past would involve a claim that the seeming 'It was' is not what it appears to be, or that it can somehow be transformed into an 'It will be' which is yet to be determined.

Who is the winner in this competition? Apparently the past, given that the present moment passes away and enters the past, requiring one to say of it: 'It was'. The same thing occurs in turn to the moment that follows and is then the present one. Martin Heidegger argues that Zarathustra's assertion that the object of the will's anger is 'time and its "It was"' identifies past time as dominating time as a whole. It does not just single out one aspect of time, but rather specifies passing away its as its overall character. As Heidegger puts it, the word 'and' in that phrase does not mean 'and also', but rather 'and that means'.[17] However, the past can never claim to have completely achieved its aim of taking over the future. Its success is only on the disputed ground of the moment. Hence, future time is not eliminated or replaced by past time. On the contrary, the two lanes are still there and in the same relation to each other when another moment replaces the one that has passed away. Their conflict can never be settled once and for all: its perpetual re-enactment is just the passing of time. Thus in the claim that 'All is empty, all is the same, all has been' the past is running ahead of the present moment, and laying claim to the entire future in advance. This is a threat rather than a *fait accompli,* but no less unsettling for that reason.

The contest in the gateway is foreshadowed by Zarathustra in his discourse 'On Redemption', where he identifies the same situation when he says that 'the will cannot will backwards' and, a few lines further on, that 'time does not run backward'. If these mean the same thing, then for time to run backward would be for the will to will backward. What if this were to happen? If the future were victorious in its struggle against the past at the gateway 'Moment', then the will to create would find itself in another one-sided predicament, no better than its existing situation. Ofelia Schutte touches on this idea by suggesting that 'the will might still be stuck in a feeling of revenge against the passing of time, only in this second, hypothetical case, the revenge would hold against the passing of time from future to past in an irreversible direction, something that is counter-intuitive to our usual experience of time but nevertheless conceivable'.[18] We can only conclude that this self-perpetuating contradiction is itself the problem. When the conflict in becoming is enframed as a fixed opposition, a realization of its creative potency becomes impossible.

Zarathustra's First Challenge

Zarathustra asks, 'But whoever would follow one of them, on and on, farther and farther – do you believe, dwarf, that these ways contradict each other eternally?' Having said that this contradiction takes place at the gateway 'Moment', he is now asking about the situation along the lanes: Is there the same conflict everywhere? There is almost a break in the middle of the utterance, and a change of direction, as if one question were left unfinished, and another asked in its place. The first question would be not about the ways, but about the person who follows one of them. Zarathustra changes his mind, and it is arguable that he makes a tactical error, because asking about the paths allows an evasive answer, as we shall see.

We need to look at both questions, starting with the unfinished one. Zarathustra speaks of following *one* of the lanes 'on and on, farther and farther'.

But is that possible? Can we assume an ability to move in either direction along the lanes? I think the answer is no, but there are some ambiguities here. Consider Zarathustra's further assertion that 'From this gateway, Moment, a long, eternal lane runs (*läuft*) backward.' There motion seems to be attributed to the lane itself and, similarly, one supposes that the other lane would run forward from the gateway. Yet the lanes are said to come together there. Unless both of these wordings are taken as metaphors within metaphors – that is, as metaphors within the figure of the gateway and lanes – inconsistent images will result. In fact, these expressions are common idioms for spatial relations. They can readily be replaced with language that has no connotation of movement, as in the statement that the lane running backward 'lasts (*währt*) an eternity'. This is all that its 'running back' means. It is also what is meant by saying that nobody has reached an end of either lane. That too sounds as if it refers to actual motion on the lanes, but this cannot be right, for there is good reason to suppose that motion can proceed in only one direction. So I suggest that when Zarathustra speaks of following one of the lanes, all he means is directing one's attention towards various locations in past or future time. That is clearly what is meant when he says, later in the chapter, 'My thought ran back' – to an experience of his distant childhood. The same is true of counting forward or backward from the present. Nietzsche's argument against a finitude of past time is that doing this is all that is required to grasp past time as infinite in extent: 'Nothing can prevent me from counting backward from this moment and saying "I shall never come to an end": just as I can count forward from the same moment into the infinite.'[19] The finitist's observation that movement in time must always be in a forward direction (that is, from earlier to later positions) is true, but irrelevant. All this is really about what takes place in the gateway, not out on the lanes. We have not yet found a need to think of travel occurring on the lanes themselves.

A puzzle is raised by Joan Stambaugh about the gateway and lanes. She thinks that Nietzsche is inconsistent in treating the moment in two ways: as the starting point *from* which one can go indefinitely far on either lane, and as a gateway *towards* which motion must presumably occur. On the first view, the moment is always the beginning, never the end – as Nietzsche puts it, the head that is not to be confused with the tail. A spatial model, however, makes no allowance for the directionality of time. Stambaugh writes:

> If past and future had the character of lanes, nothing would prevent me from coming and going on them, for they would lie stretched out before me there like every lane. And if the moment had the character of a stationary gateway, I could always come back to it from whatever direction I choose.[20]

This puzzle cannot be resolved, it seems to me, in terms of gateway and lanes alone. As it stands, that image is inevitably symmetrical, with lanes coming together at the gateway and a figurative motion of the observer's gaze away from the gateway in both directions. Only the contradiction asserted by Zarathustra has the potential to make this model anything other than a static one. The directionality of time requires a further element, and this will involve movement on the lanes.

Let us now return to the riddle itself, since only its opening part has been considered so far. What Zarathustra asked was whether these ways 'contradict each other eternally'. He is asking about all the locations on one or other of the lanes. Do the lanes still contradict each other there? This is a puzzling question, for it seems evident that as soon as we depart from the gateway that makes the lanes separate from each other, there is only one lane, and in that case there cannot be any such conflict. Hence, the answer must be that the ways do not contradict each other eternally, but only at the present moment. On the other hand, it may be argued that the situation at a given place can be grasped only by someone who is there, not by a more or less distant onlooker. On that assumption, it seems that whatever location one picks is a temporal point of view, a place from which an observer looks forward to the future and backward to the past, and as such it must be a gateway that stands between two lanes. In other words, every location on either lane has past time on one side and future time on the other – time that is past and future in relation to that position. Thus, the same situation must exist at every place along either lane. We may follow either as far as we like, but any point we arrive at will be one at which two lanes come together. Hence, one can conclude, the answer to Zarathustra's question must be that the two lanes do contradict each other eternally, assuming that they do so at the 'moment'.

Given this impasse, it is not enough just to say that the lanes represent past and future time. We need to know *how* they do that, and for this an approach is required that addresses what occurs on the lanes. That will mean considering actual movement along the lanes, and even within the gateway. This subject has not been mentioned by Zarathustra, but it will be central in his further interaction with the spirit of gravity. To see why, let us reconsider the metaphorical situation. Again, without the gateway there would not be two lanes, just as without a present there would be no past or future. In that case, would there be any time at all? A single lane would symbolize absolute becoming. Whether it can be pictured is another matter. Perhaps a time incapable of being grasped by our minds amounts to much the same thing as timelessness. Would not a pure flow reduce our existence to a becoming without a before or after, a vanishing present of the kind indicated by the Heraclitean philosopher who can communicate only through action, not representation?

Any interpretation of becoming that goes beyond this involves, first and foremost, distinguishing the temporal position from the event occurring there. In metaphorical terms, this is the distinction between a way and those who travel on it. A gateway and lanes cannot by themselves represent temporality, for there is nothing in that image corresponding to the original and essential element of becoming. A purely perspectival approach cannot solve this problem, as long as the distinction it makes between past and future is like the contrast between left and right in a spatial perspective. That logical indifference is evident in Zarathustra's first question, despite the reader's natural but premature assumption that one lane represents future time and the other past time. For any further contrast, the directionality of becoming is needed. That requires a new feature of the metaphor, not present in Zarathustra's opening exchange with the spirit of gravity but stated in his new challenge, which begins with these words: '"Behold", I continued, "this moment! From this gateway, Moment, a long eternal lane runs *backward*: behind us lies an eternity. Must not whatever *can* run have run on this lane before?"'

The Temporality of the Lanes

Instead of the two opposing lanes, we are now concerned with those who walk or run on them (either word may be used to translate the German *laufen*). The motion of travellers is, in fact, what confirms the temporal meaning of the model. It may be objected that it does this by eliminating the metaphorical nature of the image, since time is now actually present within the picture. That shows the difficulty of symbolizing time. Any representation that does not *include* time may seem inadequate, because of the need to make a contrast between time and another continuum, such as space. It seems that only becoming can symbolize becoming – or perhaps that becoming cannot be symbolized at all. When those who run on the lanes are introduced into the image of gateway and lanes, it becomes a *kinematic* model, symbolizing the passing of time as the continual motion of travellers along the lanes and, as seems to be implied, through the gateway.

The claim made in Zarathustra's question, that whatever can run must do so, applies to both lanes taken together. In this regard, they are treated as indifferent, just as in his earlier observation that 'no one has yet followed either to its end'. Conflict or contradiction between them is disregarded, and no more is heard of it. Their extension is still emphasized; both lanes are endless, and so movement on them proceeds from eternity to eternity. In one way, however, the lanes differ from each other. Those on one lane are moving towards the gateway, whereas those on the other are moving away from it. Accordingly, the gateway is not symmetrical. It has two faces, and so do those who run on the lanes, according to a notebook entry from this period: 'Everything has two faces: one passing away, the other coming to be.'[21] The observer from the gateway will see one or other face of the travellers, depending on whether they are approaching or departing, and this difference corresponds to the faces of the gateway itself.

In which direction, or directions, do those on the lanes run? To answer this we need to think further. There are important philosophical debates over time – now lasting a century – arising from J. M. E. McTaggart's argument against the reality of time, published first in 1908 and in an extended version in his 1927 book *The Nature of Existence*.[22] Crucial to McTaggart's approach is a distinction between two different ways of ordering events, according to their 'positions in time'. We can place them in one order as being past, present or future, and McTaggart gives this model a label adopted by many later writers: 'For the sake of brevity I shall speak of the series of positions running from the far past through the near past to the present, and then from the present to the near future and the far future, as the A series.'[23] The position of an event in this sequence is not permanent: any event that is present was once future and will be past. In contrast is what McTaggart calls the B-series, in which events are ordered by being earlier than some other positions and later than other positions. Their place in that series does not change: if one event is earlier than another, it can never become later, and vice versa. Every event occupies a place in both series: it is a past, present or future event, and it is before or after other events.

We need both series to have genuine time, McTaggart maintains. There are some philosophers who consider the B-series sufficient, given that it locates every event at a certain temporal location, and accounts for intervals as the distances between these. McTaggart's response is that time necessarily involves change, and that change cannot be defined without reference to past, present and future. Change must have a certain direction, and so an ordering of events is not enough, since that can be taken in either direction. What is this single direction? McTaggart points out that the passing of time can be represented in two ways. We can say that the present moves continually from earlier to later positions in the series, or else that events which were in the future move into the present and then into the past. These are really two ways of describing the same process, as he explains.

> It is very usual to contemplate time by the help of a metaphor of spatial movement. But spatial movement in which direction? The movement of time consists in the fact that later and later terms pass into the present, or – which is the same fact expressed in another way – that presentness passes to later and later terms. . . . In the first case time presents itself as a movement from future to past. In the second case it presents itself as a movement from earlier to later. And this explains why we say that events come out of the future, while we say that we ourselves move towards the future. For each man identifies himself especially with his present state, as against his future or his past, since it is the only one which he is distinctly perceiving. And this leads him to say that he is moving with the present towards later events, and as these events are now future, he says that he is moving towards the future.[24]

The last sentence clarifies and corrects a common but misleading usage. We are not really moving from the past towards the future. If I move from one place towards another, then I will reach the other place, or at least be closer to it than before. Yet as Schopenhauer pointed out, 'No man has never lived in the past, and none will ever live in the future.'[25] The truth is that we (and our 'now') are moving from earlier to later positions in time, bearing the present with us, and therefore always remaining between the past and the future. Similarly, it could be said that the events which are now in the future are moving towards us, while those in the past are moving away from us. This would at least explain the common, if equally misleading, notion that the future is 'to come', expressed in the German word *Zukunft.* In McTaggart's model, each of the two series constitutes a frame of reference which may be taken as stationary, so that the other series is imagined as moving along it. This accounts for the two idioms that he has described.

Nietzsche uses expressions that correspond to each of these versions. His general preference is to present the scheme of past, present and future as the stable framework within which movement occurs, so that its direction must be from the future to the past. As with McTaggart's series, those who run on these lanes have two positions: a place within the sequence of travellers, and a location on one or other of the lanes. They do not change their positions relative to one another: if one is earlier or later than another, that relation is permanent. However, their positions on the lanes undergo constant change: they approach the gateway from the future, pass through it and depart along the lane of the past. In describing motion towards the gateway, Zarathustra says: 'And are not all things knotted together so firmly that this moment draws after it *all* coming things?' This use of the term 'moment' is different from the earlier one. It refers not to the gateway, but to whatever is in the gateway. Only a 'thing' can be knotted together with other things and draw them along as it passes into the past lane.[26] What the moment (in this sense) draws after itself is what runs on the path of the future: thus, everything in the future is coming towards the present. On the same principle, the moment is drawn into the past by the things on that lane. So 'things' – or, more strictly, processes – move in a single direction, from future to past, like the 'events' of McTaggart's account. A further point follows from the description of these as 'knotted together', as strands in a rope are entwined with one another.[27] If that is the case, they must be moving not only in the same direction, but at the same speed. There is not only no standing but no overtaking on these thoroughfares. Thus, the travellers constitute an infinitely long train, a single procession moving with a uniform speed.

McTaggart's other model of temporality is seen in Zarathustra's claim that the greatest problem for the will is its lack of power over what has already been done: 'The will cannot will backwards.'[28] This inability is a source of frustration and anger, expressed in the impulse for revenge. The directionality of the will is identified with that of time itself: 'That time does not run backward, that is his wrath.' In a notebook entry of the same period, Nietzsche observes that, rather than going backward, 'time flows off into the future'.[29] Plato similarly says, 'is not time always moving forward?'[30] All these expressions belong to McTaggart's second model, according to which 'we ourselves move into the future'. A hint of this occurs in the depiction of gateway and lanes, when one lane is described as 'behind us' and the other as 'before us'. These terms express the experience of time from a perspectival point of reference that is taken as moving continually from earlier to later positions – expressed loosely as 'towards' or 'into' the future. As McTaggart makes clear, the two models are equivalent. Within either version there is a similarity between past and future, in that the direction of motion is the same. Whether we think of becoming as a movement of events from future to past, or of presentness from earlier to later, it has a single direction.

We can now see why the image of a single lane could not symbolize time. In order to have any sense of becoming, we need more than a distinction between locations in time and the events or state of affairs to be found at those locations, for that contrast can be made equally well with space. This 'more' is movement. With a single lane and no gateway, motion is impossible to define, so some further condition is required. We tell that things are moving by having a point of reference. In the simplest case of one dimension, motion will be 'towards' or 'away from' this position. Without such a landmark, there is no way to identify movement. That is why the condition of absolute becoming is indistinguishable from one of stasis. As Nietzsche puts it, without succession and coexistence, a continuum of force would appear to us 'to be one, at rest, changeless, not a becoming, without time and space'.[31] This is the reason for introducing travellers into the model of gateway and lanes: it is to represent the passage of time by means of succession, and so to avoid the quandary identified by Nietzsche as a consequence of the 'human antithesis'.

What is it like to be out on one of the lanes, as distinct from looking out from the gateway? Here one is faced with a dilemma. One thing we know is that those who are on the lanes are walking or running on them. This is what makes them distinct from the lanes themselves, and makes the image one of time and becoming. Are they on a single lane that would symbolize absolute becoming? There would be no past or future for anyone in this situation, and Zarathustra's question about the gateway would make no sense. On the other hand, if they can be said to look forward to the future and backward to the past, then they are doing that from the standpoint of some present moment, which must be a gateway at which past and future come together.

Some scholars do take this view. For example, Eugen Fink argues that the lanes must be understood in terms of the 'moment', just as much as the gateway. He writes, 'It becomes clear that Zarathustra approaches time immanently. Time is conceived as a sequence of moments (*Jetzten*). A given moment is preceded by an infinite sequence of moments in the past and followed by a similarly infinite sequence of future moments.'[32] Claudia Crawford writes, 'If the gateway is the "moment" then the lanes are composed of "moments".'[33] On this interpretation, it could perhaps be said that there are no lanes, only gateways. That must be wrong, since a gateway requires lanes to be there at all, and yet Fink is right in saying that Zarathustra's view of time is an immanent one. Both sides of the dilemma, it may be argued, make the mistake of treating those on the lanes on the same basis as the observer in the gateway. The perspectival standpoint then loses its determining status and is replaced by the kind of objective picture that Nietzsche rejects. A stronger way of expressing this would be in terms of actuality. Schopenhauer's claim that the present is the only real form of time is identified by Nietzsche with the doctrine of becoming. And Zarathustra in fact refers to travellers on the lanes only in general terms: whatever can run must run there, and all that is to come is drawn after the present moment. This is all he wants to say about what is on the lanes.

The Temporality of the Gateway

Returning to the gateway may help us to get a more complete picture. Is there a kind of motion correlated with the gateway? If travellers on the lanes pass through the gateway, then they must be walking or running as they do so. Yet there are hidden puzzles in this ready answer, for any divorce between motion and extension gives rise to conceptual problems uncovered by Zeno and still under debate. How can an arrow move in an indivisible moment? It simply occupies its position: there is nowhere else for it to move to. Motion within the gateway is problematical for much the same reason.

Still, there is another option: to take it as a place where one stops, or at least pauses. Before considering this idea, though, we need to avoid a possible confusion. The gateway is mentioned first in Zarathustra's statement that 'there was a gateway just where we had stopped'. Yet it was not *because* he arrived at a gateway that Zarathustra came to a halt.[34] Before stopping he had been climbing a mountain path, not travelling on either of the lanes, and gateways are not common on steep and rocky paths. This event has nothing to do with movement on the lanes or in the gateway. The path is where starting and stopping occur continually, so that a pause is always possible. That discontinuity is liable to constitute a moment on condition that the tension experienced as between forces in two directions, up and down, is redefined as one between forward and backward (or vice versa) and so comes to indicate the collision of future and past as a whole. This is a shift from one interpretation of the conflict of force in becoming to another. In Nietzsche's text, the transition is signalled by the division between the two sections of 'On the Vision and Riddle'.[35]

On his own account, Zarathustra stops because he makes a courageous decision to confront and challenge his enemy: 'This courage finally bade me stand still and speak: "Dwarf! It is you or I!"' Courage is needed to bring on any confrontation from which only one antagonist will emerge. What does that have to do with stopping? The riddle that constitutes the decisive contest can be asked *only* at the gateway, because that is what it is about. Thus, the location of the gateway at this place is no coincidence. The circumstance of the challenge and its content go together.

We come back, then, to the question about the gateway. Is the gateway there only for the traveller who pauses? Its role as a spot from which an observer may survey the lanes is consistent with being stationary rather than in motion. From this standpoint, one can look forward or backward, so that the present moment can figure as the beginning of future time or the end of past time. The word 'standpoint' already suggests a standstill, that is, a state of rest, although not necessarily a permanent one. The notion of the present moment as dividing past and future time in the way that a point divides a spatial line seems to support this idea. Aristotle observes that giving the moment two roles, as the end of one part of time and the beginning of another, means counting it twice – for which, he says, 'a pause is necessary'.[36] Here he is assuming that any particular motion has a beginning and an end, each involving a transition between rest and motion. This idea however depends on the notion of a pause, a momentary disturbance in the process of change. Yet for a Heraclitean like Nietzsche there is only a continuous flow of becoming, so it is hard to see how any such pause can occur. Worse still, he has argued that any pause whatever would result in a permanent standstill. Assuming no intervention from outside, there would be nothing to bring about a renewal of change. So it is hard to see how such a concept of the moment can be justified within the doctrine of absolute becoming which he so often endorses. The same comment applies to his model of gateway and lanes. The travellers who run on the lanes do not alternate between motion and rest, or even make a transition from one motion to another. How can they stop, even for a moment? Here is another puzzle for our argument.

One solution is to take the moment as something timeless. In his reading of this text, Robert Gooding-Williams calls the moment 'a perpetual or eternal "now"' and concludes that the lanes 'stretch the presence of the present moment eternally forwards and eternally backwards into an eternal *nunc stans*'.[37] He supports this by saying that 'For Zarathustra, all places present on the paths of the past and future are present simultaneously. Thus, each of these places is present at the same time as the place he himself occupies.'[38] It seems to me that his argument rests on a misunderstanding of Nietzsche's use of a spatial metaphor for time. Instead of taking simultaneity to be represented within the model by sameness of spatial location, he attributes it to the model itself, as if it were an example rather than a metaphor. His case now looks like a refusal to represent time as space, or perhaps as anything other then time. Well, could one take that position and provide some supporting argument? It is hard to see why *parts* of time should not be represented spatially, since their mathematical properties are those of a line. It is the forms of time, and their links with becoming, that present issues there. Nietzsche implicitly concedes that the gateway and lanes, taken by themselves, cannot represent time, since they leave out something essential. He deals with this by introducing a further element: the travellers on the lanes. Until that feature is considered, it is too soon to say that the moment must be seen as timeless because of the model's inability to distinguish past and future time from the 'now'.

Even if the gateway 'Moment' can be identified as a place where a pause occurs, enabling it to act as a perspectival standpoint, it must also be where the passage from one lane to another occurs. I have taken it that the direction of this movement is from the lane representing future time to the past lane. Some commentators raise another question: does Zarathustra himself pass through the gateway? Nietzsche's text gives no answer, but the idea's appeal to many readers is evident. For one thing, the impending disappearance of the spirit of gravity would be explained if, as Stanley Rosen puts it, 'Zarathustra does not persuade the dwarf to pass through the gateway'.[39] Rosen adds that Zarathustra himself has difficulty in making that transition, since 'the hardest step of all is to enter into the future'. According to Paul Loeb, Zarathustra certainly does pass from the past to the future lane.[40] He cites as evidence a shift in Zarathustra's wording, which refers to the lane ahead of him first as '*that* (*jener*) other lane' and then, shortly after, as '*this* (*dieser*) long horrible lane'. Yet Zarathustra has earlier called it 'that long lane out there' and 'this long lane out there'. Either Nietzsche's use of 'this' and 'that' is not so precise, or Zarathustra is somehow going back and forth between lanes. As for the dwarf, Zarathustra has threatened to leave him 'where you are crouching' – that is, on the height, not on one of the lanes.

There is an obvious problem with Zarathustra's passing from past to future: he would be going against the traffic of 'all coming things', that is, the train of travellers moving from the future into the past. However, both Rosen and Loeb are thinking of the gateway not as a generalized 'now' but as a unique event in Zarathustra's personal journey, a transition which is also a personal transformation. For Rosen, it marks a shift from what he calls 'cosmological time' to 'historical time'.[41] For Loeb, it is no less than the moment of Zarathustra's death, coinciding with his rebirth into a new, but identical, life. Both of these ideas raise issues lying outside this discussion. But both also omit some of what is said by Zarathustra about the gateway and lanes. Hence, Rosen is forced to offer separate but concurrent readings: in one sense this is the temporal structure common to everything, but in another sense, it belongs to an individual life. His explanation of the first sense tries to resolve the puzzle about motion in the gateway:

> We enter the gateway at the moment, which corresponds to the present. Having so entered, we must walk simultaneously forward into the future and backward into the past. And in so doing, we remain continuously within the gateway, or what is for us the present moment. At the moment of death, we can no longer walk forward but are as it were conveyed ever farther backward into the past.[42]

The trouble with this account is its conflation of the observer's perspectival outlook on the two lanes with the traveller's motion, which can hardly be in two directions at once. In the background are the two models of time: one involving a moving 'now' with stationary events, and in which time may be described as moving forward, and the other involving a stationary 'now' (symbolized by a fixed gateway) and a train of events that move towards and through it, going from the future into the past. Rosen wants to explain how the conscious observer is always in the present, and yet, if identified with the events of his or her life, is also in the past and future, and eventually wholly in the past. It is this appeal to both models at once that gives rise to the awkward mixture of themes in his description. But I think there is another issue here as well. As with the pathway, a visual image does not tell the full story. We need to grasp what is happening in another vocabulary: the language of forces, their interactions and conflicts, touched on in relation to the gateway but not yet explored. This is just the dimension missing in most readings of 'On the Vision and Riddle'.

The Dynamics of the Gateway

The symbolization of becoming as motion is a translation into what Nietzsche calls 'the sign language of sight and touch'.[43] Yet these are two distinct languages. One is the visual imagery that constructs a kinematic picture of becoming, while the other's vocabulary centres on the concept of force. This is just as important in the metaphorical structure of 'On the Vision and Riddle'. The two lanes come into confrontation in the gateway, pushing forcefully against each other. Their infinite extent is not a factor here: what characterizes force is its intensive magnitude, as well as its direction. Further, travellers on the lanes play no part in the interaction of forces within the gateway. When they are introduced, there is no suggestion that they are influenced by these forces. Each traveller is held fast in a different way, in that each maintains a fixed position in relation to the others, but that is explained in terms of their mutual 'entanglement', not by forces acting on them from outside. So, this notion of interacting forces is related to the gateway rather than the lanes.

The contrast can be taken further. A perspective is about what is observed from a certain standpoint. Hence, the gateway is the origin from which the lanes extend. In effect, they are the projections of the observer who is located there and whose survey, Nietzsche thinks, is sufficient to establish that both lines are infinite. For that reason, the restriction of actual travel along the lanes to one direction (a crucial premise in the *kalām* argument against an infinitude of past time) is discounted as an objection. Moreover, perspective is a visual and spatial metaphor, corresponding to a contemplative attitude. The perspectival approach implies detachment and distance – the longer the better, which is why, as Nietzsche reminds us, 'the devil has the broadest perspectives for God'.[44] Force, in contrast, is not seen but rather felt, as weight, tension or pressure. The dynamic approach is a description of struggle and overcoming, one that emphasizes processes rather than products. It bears on the here and now, not on what is distant and separate.

These conceptualizations of time proceed in different directions. In either case, the gateway stands between two lanes. In the perspectival account, the point of view is the source of meaning, for the distinction between past and future depends on the standpoint of a present from which they are defined. In the dynamic account, the moment must be seen as a product, resulting from the convergence of past and future. That collision is just what makes the gateway possible. In contrast, the kinematic and perspectival approaches take the gateway and lanes as given, and use their metaphorical vocabularies to develop the model's further implications, as we have seen. They do not attempt to account for the presence of the gateway in the first place: only an alternative that starts from a conflict of force can address this question.

What, then, are these forces that confront each other in the gateway? This is a difficult problem, and we need to look around for clues. Hannah Arendt contributes a pertinent discussion of a related image in *The Life of the Mind.* She is concerned with what she calls 'thinking', especially in its relation to time.[45] To illustrate this theme she takes up a short parable of Franz Kafka, one of a collection labelled 'He' by his editors:

> He has two antagonists: the first presses him from behind, from his origin. The second blocks the road in front of him. He gives battle to both. Actually, the first supports him in his fight with the second, for he wants to push him forward, and in the same way the second supports him in his fight with the first, since he drives him back.[46]

The individual here is caught in the middle, trapped and held fast by forces which together prevent any movement, forward or backward. But, Kafka continues, this person may also have his own plans, intentions and wishes, which could include an aspiration to 'jump out of the fighting line and be promoted, on account of his experience in fighting, to the position of umpire over his antagonists in their fight with each other'.

Arendt notes in passing that Nietzsche's allegory of gateway and lanes is 'a curiously related story',[47] but her analysis is based on the Kafka fragment, where the collision of pressures is oppressive and disabling for the anonymous 'He'. Arendt takes this to describe the typical situation of the conscious person in relation to past and future time, and believes that the prospect of escape from the struggle is offered by an activity not confined within time. This is thinking, as she understands it. In her version, the collision of past and future leads to a productive outcome, a force acting in a new direction, that of thinking. Arendt emphasizes that she does not have in mind any flight to a region of thought 'situated beyond and above the world and human time'.[48] The thinker remains situated between past and future, but in a 'present which is timeless', and which 'gathers the absent tenses, the not-yet and the no-more, into its own presence'.[49] In Chapter 8 I will be looking into Nietzschean ideas about a reconciliation of past and future that are closely related to Arendt's conception, and so will return to this theme there. For the time being, I am concerned with understanding the situation rather than considering a solution.

My interpretation of Nietzsche postulates a productive role for the conflict of force in the moment, but regards it as constitutive for temporality itself, or at least this form of temporality. As Gerard Visser puts it, 'The moment is the time location (*Zeitraum*) of accidentality.'[50] It is where the interaction of the creative impulses occurs, the nearest thing to a discontinuity that is available for absolute becoming. This occurrence is what makes possible a temporal interpretation of becoming that starts from a present moment and a corresponding past and future time. The stream of becoming does not cease, but its smooth flow is disturbed by its own inner turbulence, grasped by us (since our thinking turns differences into oppositions) as the collision of forces whose outcome is the fleeting stasis of the moment, the 'pause' that makes it a perspectival standpoint.

What can we learn from the dynamics of the gateway? For Nietzsche, a conflict of force is the most basic level of experience. There is a continual play of forces within our bodies – in fact, on Nietzsche's view, this is just what the body is. However, we interpret sensed forces in various ways. For example, we take them to be the qualities of some more or less permanent object, or of non-material centres of force: in either case, forces are given temporal properties. Thus, we try to explain the felt conflict in terms of the opposition between 'forward' and 'backward' directions. We *interpret* a conflict as between the past and future, or between an earlier period of time and a later one, such as yesterday and today. In both cases, the dynamic account is the primary key to an interpretation of becoming as time.

One last remark before returning to 'On the Vision and Riddle' and the dwarf's response to Zarathustra's challenge. In identifying the gateway as a place of momentary rest, we have made progress in understanding Nietzsche's statement: 'He who would learn to fly one day must first learn to stand and walk and run and climb and dance.' Standing is what takes place in the gateway, walking and running on the two lanes, climbing on the mountain path. What about dancing and flying? These forms of motion, and the way or ways corresponding to them, remain to be identified. That will come later.

Chapter 6

Linear and Circular Time

None of these issues are uncovered in Nietzsche's text. Instead, the dialogue shifts abruptly. Zarathustra asks whether the ways contradict each other eternally, and the spirit of gravity immediately responds. '"All that is straight lies", the dwarf murmured contemptuously. "All truth is crooked; time itself is a circle."' In response Zarathustra angrily accuses the dwarf of making things 'too easy' for himself. It seems that the answer is wrong, or not a genuine answer at all. Yet on the face of it, the answer is straightforward, even if it does not address the question directly. Time, it says, is a circle. What that implies is that the two lanes are not two lanes after all. Everything that seems to be in the past is also in the future, though in a way that is hidden from our point of view, from which we cannot perceive the meeting of the two lanes. Conversely, everything in the future is also in the past. In much the same way, one might say that since the earth is round, whatever lies to the East may also be said to lie to the West, and vice versa. What does this say about the conflict between past and future? There can be none on this hypothesis, because there is no difference between past and future. If they seem to be in conflict at the 'moment', that must be an illusion. The dwarf's answer thus introduces a distinction between appearance and reality into our experience of temporality. The truth is a reality which is not accessible from the gateway: it is that time is curved rather than straight, and in fact a complete circle.

Is the dwarf answering the question at all? What it calls for is a 'Yes' or 'No' response, and he refuses to give either. Yet his opening statement that 'All that is straight lies' is well directed. Nietzsche writes, 'Formula of our happiness: a Yes, a No, a straight line, a *goal*.'[1] A straight answer ('Yes' or 'No') would correspond to a straight line: that is, it would assume that the two lanes are straight, since only in that case could they be said to contradict each other and collide head to head. Hence, as with the upward and downward forces experienced on the mountain path, linear directions are presupposed by Zarathustra's description, even if he does not say so. What the dwarf proposes is a completely different model, symbolized by a circle.

To understand the ideas advanced by the spirit of gravity in these few and cryptic remarks, we must know that Nietzsche had encountered the claim that time is like a circle in one contemporary philosophical work. The writer was

Gustav Teichmüller, whose major work *Die wirkliche und die scheinbare Welt*, published in 1882, presents a metaphysical system that bears a close relation to Plato or even Parmenides. Nietzsche's study of this book is evident in his use of its key terms, such as 'perspective', 'projection' and 'semiotic', from this time onwards. Mentioning the idea of circular time in passing, Teichmüller refers readers to his earlier work *Darwinismus und Philosophie* for a fuller explanation.[2] The idea appears there in a discussion of origins, designed to support a rebuttal of Darwinism. Must we think of the world as having a beginning, and how is such a beginning to be understood? For Teichmüller this is a misconceived question. In looking for the causes of things in the past, we forget that they are just as much conditioned by what lies in the present and the future.[3] Nature is a system whose laws have nothing to do with time, a totality in which every element is conditioned by every other element.

This leads him to assert that 'time behaves like a circle (*Die Zeit sich verhält, wie ein Kreis*)'.[4] A circle has no beginning or end. Or rather, any point on it can serve as a beginning from which one can travel in either direction and return to the same point, which is thus as much an end as a beginning. It is the same with every moment of time. Teichmüller provides no very convincing argument for his claim, but he does reply to the objection that time at least *appears* to be a straight line. This is, he says, a perspectival illusion arising from the limitation of our field of view. In much the same way, we ordinarily perceive the earth as a flat surface, and yet have been persuaded by scientific understanding that it is really a sphere. It is true that the idea of time which occurs most readily is a linear one, but this too is an illusion. Fortunately we can rely on reason to overcome the error, and so avoid the insoluble problem of finding a beginning to the process of becoming. Teichmüller writes:

> It turns out that this question can be raised at all only if the person who asks it tries to survey the totality from his particular standpoint, and thus adopts a mere perspectival picture, in which the crooked must appear straight and the straight appear crooked. In fact, the totality in itself, i.e., not regarded perspectivally, has neither beginning nor end, but constitutes a timeless unity.[5]

Teichmüller's perspectivism is closely bound up with a denial of the reality of time. The finitude of our standpoint gives rise to perspectival illusions. For example, we think that what has happened in the past cannot be made not to have happened, whereas the future is, so to speak, not yet there. But the distinction between what has happened and what has not happened is only a perspectival rule, even if it applies to every finite standpoint.

> Taken absolutely, one can and must regard the whole sequence of phenomena as complete, so that, to put it paradoxically, the future has already occurred, and past and future are simultaneous. Such an intuition would, however, be possible only for the absolute consciousness that we attribute to

> God; for the individual real subject which, with its real activity, belongs within the total order as an individual member, what is given in intuition is always a perspectival ordering of time.[6]

Our notion of free will, Teichmüller explains, expresses this 'openness' of the future.[7] But if the world is grasped as a timeless totality, no such notion need arise.

Teichmüller describes the usual view of time as 'a mere perspectival picture, in which the crooked must appear straight and the straight appear crooked'.[8] Nietzsche repudiates this position: 'God is a thought that makes crooked all that is straight, and makes turn whatever stands. How? Should time be gone, and all that is impermanent a mere lie? To think this is a dizzy whirl for human bones, and a vomit for the stomach: verily, I call it the turning sickness to conjecture thus.'[9] That sounds like a strong condemnation, and the same could be said about the way Zarathustra responds to the dwarf's reply to his question. Yet it would be hasty to suppose that Nietzsche had a low opinion of Teichmüller or his philosophy. He agrees that empirical knowledge is always directed towards 'perspectival appearance' rather than reality, but not with Teichmüller's further belief in the ability of reason to rise above this world of appearance and grasp reality as it is in itself.

The Dwarf's Reply

This dispute over appearance and reality figures in *Thus Spoke Zarathustra,* and nowhere more than in 'On the Vision and Riddle'. The dwarf separates appearance from reality, and condemns appearance as a lie. His answer is contemptuous – of *appearance.* A reader might suppose that the object of his contempt is Zarathustra, or Zarathustra's question. What counts against this, not for it, is Zarathustra's angry response. It would be out of character for him to be angry at contempt directed towards himself, especially from a source for which he has no respect. His earlier encounter with the bystanders of the marketplace has established this point, since their mocking response to his discourse on the *Übermensch* saddened rather than exasperated him.[10] Zarathustra is not concerned for his own reputation, but he is a champion of the validity of appearance, and so is angered by the claim that appearance is a lie. In his view, it is the so-called real world that is a lie.

A number of writers argue that Zarathustra's view of time is indeed the one stated by the dwarf, or else that, whether or not Zarathustra accepts it, the dwarf's answer is essentially correct. If that is the case, then Zarathustra has something to learn from the dwarf, perhaps in line with the saying that 'A dwarf placed on the shoulders of a giant sees further than the giant himself.' One must still wonder why Zarathustra is angry with the dwarf. A possible answer to this lies in his remark that 'Not every word belongs in every mouth.'[11] What would be true if spoken by Zarathustra himself is false coming from this speaker, who is not entitled to a knowledge that has not been acquired and grasped on his own account, just as he is not entitled to be on a mountain height that is beyond the reach of his own efforts. Zarathustra warns his 'higher men' against willing 'beyond their capacity'.

> If you would go high, use your own legs. Do not let yourself be *carried* up: do not sit on the backs and heads of others. But you mounted a horse? You are now riding quickly up to your goal? All right, my friend! But your lame foot is sitting on the horse too. When you reach your goal, when you jump off your horse – on your very *height*, you higher man, you will stumble.[12]

The dwarf does stumble. Unable to stand by himself on the height, he can only crouch. He has tried to make things 'too easy' for himself. Hence, one might conclude, Zarathustra's anger at his contemptuous reply to the question. Whether what the dwarf says is true or false is another matter.

This account is cogent as far as it goes, but it leaves out the issue that is crucial to this debate: the question of appearance and reality. When that becomes explicit, it is plain that the spirit of gravity's understanding of time as mere semblance cannot provide an acceptable answer to Zarathustra's challenge. The dwarf's intention is not to solve the problem, but to dissolve it. If the question contains a false assumption, it cannot be answered on its own terms. The dwarf denies that the situation is really as Zarathustra has described it. The contradiction between past and future, occurring either at the gateway or anywhere else, is an illusion. Since time is a closed circle, the time on one side of the moment is the same as the time on the other. Hence, if 'It was' and 'It will be' are incompatible, one or the other must have been wrongly attributed to what is, in reality, time as a whole. If these predicates do not just exclude each other but jointly exhaust the possibilities for any characterization of time, then we must say that time as a whole is either an 'It was' or an 'It will be'. Accordingly, there must be either no 'It was' or no 'It will be', despite appearances. But which of these alternatives is the right one? The presumption that past and future meet at their distant ends raises the same issue as the one embodied in the confrontation between lanes at the gateway 'Moment'. To this problem, the dwarf's answer offers no answer. His failure to resolve the contradiction reveals the vacuity of his response, even taken on its own terms.

The spirit of gravity's stratagem is an attempt to go beyond a perspectival picture of time whose standpoint is the present moment. The alternative is supposed to be a knowledge which does not locate itself within time, and for which time, with its distinctions between past, present and future, is mere appearance. This is the view that time is a circle. Paul Loeb suggests that 'Although the dwarf mouths the words "time itself is a circle," he does so contemptuously (*verächtlich*) and does not actually believe that these words describe true reality.'[13] This implies that the dwarf's contempt is directed towards his own words – unlikely behaviour for one who is 'serious, thorough, profound and solemn'.[14] Not many further clues to the dwarf's world view are given. We know that he is the spirit of gravity, 'through whom all things fall'. This does not by itself entail an opinion about reality, and yet it is significant that his pronouncements are typically universal ones: 'every stone that is thrown must fall', 'All that is straight lies' and 'All truth is crooked; time itself is a circle.' Similarly, in a later chapter, he is quoted as saying: 'Good for all, evil for all', without any occasion for irony being suggested.[15] There is a consistency here. Just as the spirit of gravity rejects the validity of a perspectival view of time, so he rejects a perspectival view of value, and insists on one universal and impersonal standard of good and evil. For this reason, Laurence Lampert calls him a representative of Socratic rationalism.[16]

Zarathustra in contrast speaks in terms of what is particular and immediate, making no claim to grasp the world as a whole. His earlier utterance began: 'Behold this gateway, dwarf', and in replying now to the dwarf he brings the subject back with a similar command: 'Behold this moment!' His question concerned the paths defined by the standpoint of the gateway 'Moment'; the dwarf does not answer this question, or even attempt to answer it, since he makes no mention of either gateway or lanes in his response. If Zarathustra's question makes sense only to someone who has real experience of the gateway, then we need to consider why this is lacking and even impossible for the spirit of gravity.

Martin Heidegger's reading of the passage is that when Zarathustra says, 'Behold this gateway!' he is inviting the dwarf to *enter* the gateway, and that by being afraid to do so, the dwarf fails the test set for him.[17] He is unable to answer Zarathustra's questions because they call upon an experience which is available only to someone inside the gateway. Heidegger tells us that 'To see the Moment means to stand in it. But the dwarf keeps to the outside, perches on the periphery.' Yet the *worst* place to see any gateway is always inside it. The best views are from outside, and preferably at some distance. Hence, an instruction to look at the gateway is actually inconsistent with standing in it. Going further, one could argue against Heidegger that the dwarf cannot see the gateway just because he is *within* it. After all, where else could the dwarf possibly be? He is lame, and so cannot walk or run on either of the lanes. In any case, it does not go without saying that travellers on the lanes see the gateway, a point I shall return to later. But standing within the gateway, one certainly sees the lanes, the subject of the debate. Still, I think that Heidegger is correct in saying that the dwarf does not enter the gateway. One might object that Zarathustra later refers to 'I and you in [*in*, not *an*] the gateway, whispering together'. Perhaps this is a minor slip on Nietzsche's part. Having come down from Zarathustra's shoulder, the dwarf crouches on a stone. That means he has not left the mountain path, for there are no stones on the lanes, which are clearly broad and busy thoroughfares, or in the gateway, which is an open space.

Heidegger's point, then, may be that one can experience the moment as a *collision* of past and future only by locating oneself where they come together, and feeling the two opposing forces as they press on either side. This is very plausible. The dwarf shows no appreciation of the meeting of forces in the gateway. His words and behaviour show his awareness of the forces in action on the mountain path: the upward-striving force that he shares in as a passenger on a climber's shoulder, and the downward-drawing force that coerces him into crouching helplessly once this support is gone. In his denial of the collision occurring within the gateway, he imagines a harmonious reconciliation of past and future through a circular unity of time which makes them essentially the same. But this is false to the experience of anyone standing within the gateway.

I said earlier that if the dwarf is unable to see the gateway, that could be because he is within it. Yet we have now rejected this idea. Could it be that the dwarf does see the gateway after all, but contemptuously dismisses both it and the lanes as mere semblance? A more straightforward reason for his not recognizing the gateway and lanes is available. This is just that the spirit of gravity is *blind*. After all, he is described not just as a dwarf but also as a mole, and moles are known for their blindness, as Nietzsche remarks in several places.[18] Further, *Daybreak* speaks of 'the blind force of gravity'.[19] The two symbols are thus linked by the common motif of blindness. At any rate, the dwarf's lack of sight explains why he makes no mention of the gateway or lanes in his response, but expresses only generalities.

It may be objected to my hypothesis that Zarathustra tells his opponent to 'Behold (*Siehe*) this gateway!' Surely an unreasonable order, one might think, if the dwarf is incapable of seeing anything. Well, yes – but then, Zarathustra has no concern for this sort of fairness. He fully intends to ask a question that he knows the dwarf cannot answer. He has declared an intention to defeat the spirit of gravity by targeting his weaknesses, and the call to 'behold' does just that, preventing any chance of a successful response. Even so, the dwarf does manage a counter-attack, and we can now understand why his reply takes the form it does. He makes a virtue of necessity by devaluing visual appearance in general, a strategy that recalls Aesop's fable of fox and grapes. It is consistent with this that his assertions use concepts that do not rely on vision to be grasped, such as the contrast between straight and curved lines. These statements need to be taken seriously because, whether right or wrong, they open up important issues about what Zarathustra means and what an adequate answer to his question would be.

Linear or Circular Time?

Discussions of the dwarf's reply tend to be overshadowed by what is soon to come, the thought of eternal return. No aspect of that has given rise to more confusion than its relation to closed or circular time. It is natural to symbolize the idea that the course of becoming is a recurring cycle by picturing it as a circle. Nietzsche too does that, especially in the dramatic passages of *Thus Spoke Zarathustra* where the doctrine is revealed. There Zarathustra describes himself as the advocate (*Fürsprecher*) of the circle.[20] Yet in his crucial confrontation with the dwarf in 'On the Vision and Riddle', the statement that 'Time itself is a circle' meets with a hostile reception, although it is not directly contradicted. Not surprisingly, there is disagreement over whether Nietzsche or Zarathustra does reject the notion of time as circular in form.

The conceptual difficulties can be seen in a common but, on closer inspection, problematical phrase, 'cyclical time'. Is time itself a process something that happens? If so, can it happen more than once? Eternal recurrence may be imagined as a circular course that is traversed again and again, but in that case, the circle symbolizes the events that make up the course of becoming, representing them as occurring before or after one another at greater or lesser distances. Talk of *motion* begins only when we think of them as coming to be and passing away, and as being in the past, present or future. All this also applies when time is symbolized as a straight line, but there the line is not supposed to be traversed more than once. The problem is to see how repeating cycles can be imagined for time itself. Is this a genuine issue, though? It may be that what is meant is a circular time with no more paradoxical consequences than linear time. A number of writers argue for that thesis as a further development of arguments for eternal recurrence. The reason given for taking the additional step is the principle of the identity of indiscernibles. They hold that moments of time which cannot be told apart by differences in their content must be taken as identical. In other words, every total state can have only one location in time, and so cannot occur more than once.

The notion of a closed time is akin to that of a finite but unbounded space, introduced into geometry by Bernhard Riemann in the nineteenth century.[21] Friedrich Ulfers and Mark Daniel Cohen take Nietzsche to be suggesting 'that time possesses a Riemannian geometry, or more exactly, the extension that is time has a positive curvature, as it is conceived by Riemannian geometry, meaning that time curves back upon itself'.[22] But does a notion of a curved time make sense? The fact that space has several dimensions is important for the concept of curvature. There is a procedure for finding out whether space is curved: by seeing whether the angles of a triangle add up to two right angles or not. That can be done on any scale, without considering space as a totality. But how could we find out whether time is curved on a local scale, let alone determine its degree of curvature? As Ulfers and Cohen point out, we are incapable of perceiving any curvature locally. The only basis for using the term is just a supposition that time is finite and closed in form. Does that mean that its curvature is constant, like a circle rather than an ellipse? On that wholly arbitrary assumption, a 'radius' could be calculated from the period of circulation, treated as a circumference. While this analogical construction is possible, it would be a pointless exercise. Better just to imagine time as a closed figure and leave it at that.

Does it make sense to speak of a closed time, though, setting aside the issue of constancy? It is sometimes argued that this idea is incoherent, because of the nature of time and, in particular, the features that distinguish it from space: change or becoming, and the distinctions between past, present and future. One such critic is Milič Čapek, who argues that those who speak of a circular time are really thinking of a linear time, but are doing so 'unconsciously' or at best semi-consciously.[23] No doubt this may be the case. Yet how we know what others are thinking unconsciously – especially people we have never met – is a difficult question. It is not likely to be resolved simply by an analysis of what they say or write, since the logical implications of a proposition are one thing, and the causal relations between conscious mental acts and subconscious ones (if there are such things) are quite another. My suggestion is that we leave this as a side issue, and concentrate on the conceptual question whether circular time has implications involving a linear time. This claim arises only when one is considering repetition: for shorter periods of time, the two models are much the same. However, if one imagines the circle as traversed more than once, there must be a time (it is argued) in which the second cycle is distinct from the first, and that will be linear rather than circular time. And without that image, there is no recurrence, but only occurrence.

In addition, Čapek objects that a closed form is inconsistent with a 'going on' that he takes to be essential to the nature of time.

> A time consisting of a single cycle obviously cannot go on; once the terminal point is attained, the world history would be at its end – unless a new cycle is postulated. If one insists that every moment is both initial and terminal in the same sense as all points on a circle are, then everything would be standing still and time, change and becoming would be completely eliminated. Zeno and modern Neo-Eleatics would be pleased – but who else?[24]

The first part of this argument seems to me mistaken. The dilemma is posed between thinking of some particular point as 'initial' and 'terminal', and thinking of every point as both 'initial' and 'terminal'. A third possibility is not considered: that no point is either initial or terminal – that every point is indifferent to these determinations, as points on a circle surely are, despite Čapek's statement to the contrary. But that is just what becoming (which I assume is the same as 'going on') implies. The failure of Čapek's argument is due to his refusal to consider becoming as a characteristic of each moment, taken by itself. Rather, becoming is thought of as, so to speak, what gets us from one point to another point, just as movement in space does or, for that matter, Nietzsche's 'motion in time'. Starting and finishing are its essential features. Hence, Čapek's talk of 'attaining'. He assumes that the completion of a particular transition is followed either by stopping or by proceeding on to a new motion that 'attains' a further point. In that case, becoming is located outside the moments, which become fixed points of reference, either initial or terminal or both. This is nothing like Nietzsche's view of becoming as the fundamental reality, encountered directly rather than inferred from the states on either side.

Several writers have speculated on the possibility of a closed B-series. George N. Schlesinger suggests that 'circularity may arise through the "now" in its voyage along the circumference of the circle formed by the B-series, reaching the same moment again and again on the completion of each cycle'.[25] Paul Loeb argues that time can be thought of as a recurring circle only if the positions in the B-series actually come into being through this movement of the 'now'. Thus, when becoming returns to a given state, it does not encounter an existing location, but creates one anew.

> Time, therefore, cannot be said to introduce any qualitative difference among recurrences, and eternal recurrence is not to be imagined as an infinite number of cosmic cycles (or spirals) succeeding each other in some absolute linear time. Instead, Nietzsche encourages us to imagine just a single finite (though unbounded) circular course (*Kreislauf*) in which is represented, not just the recurrence of all things, but also of all those moments of time that cannot exist independently of those things. Since for Nietzsche time is just a series of those moments, it follows that time itself is destroyed, re-created and repeated along with everything else. Thus, when I am re-created so as to relive my identical life, the time at which I am re-created and live my life is always exactly the same.[26]

On this view, time itself is destroyed and re-created along with events. Thus, the occurrence of an event is also the creation of its location in time. This is not such an unusual position. It is encountered in the finitist approach which rejects any actual infinitude of time and relies instead on a quasi-Aristotelian notion of a 'potential infinity' to guarantee an endless future, and perhaps past as well. It is just because one can always continue onward that time can be described as infinite. We need to note a comment made by Georg Cantor, as part of his campaign to establish an actual infinite in mathematics. He is here replying to two critics who defended a characterization of the infinite as a continually changing limit. They invoked the standard metaphor which represents the potential infinite as a kind of movement: however far we progress, it is always possible to go further. Cantor responds in the same terms by pointing out that without firm ground beneath one's feet, any such excursion becomes an uncertain business:

> Have these gentlemen completely forgotten that, apart from journeys carried out in imagination or in a dream, a firm ground or base, and a ready-made way (*ein geebneter Weg*) are, I point out, unconditionally required for safe travelling or wandering – a way that does not break off anywhere, but must be usable and remain so wherever the journey leads?[27]

Cantor concludes on a sarcastic note by wishing his critics a 'happy journey'. While he is thinking of an infinitely extended 'way', the plausibility of his argument remains if one imagines a closed path. In either case, the possibility of going further depends on having a place to go to. A version of closed time that imagines time itself as being continually created by the process of becoming faces this telling objection.

In any case, why speak of time apart from events, if they are really one and the same? Perspectivism implies a relational structure distinct from factual content, and so is inconsistent with a relativism that explains everything in terms of content, reducing space to objects or time to events. And more to the point, just what is the distinction between recurrence and simple occurrence here? The only other feature of the description given above is the circularity of the course of becoming. But is not that circularity itself just a way of expressing the fact of repetition? It is not something that we can identify apart from and, in particular, before its completion. In contrast, the circular path of an object moving in space can be identified by observation, and the object's eventual return to its starting point can be predicted in advance. This cannot be the case with the world as a whole. Any total state of things, taken by itself, is consistent with either a circular course or an infinite linear progression. Loeb's position must be therefore rest on a further claim that he makes, that we can *remember* a state of affairs as having occurred before. We shall return to this intriguing idea in the next chapter.

My conclusion is that these arguments fail to show that the concept of closed time is incoherent. As I have said, this does not mean that it can be identified with eternal recurrence. With a circular time everything happens once or, if there is a repetition of events within the circle, it is only a finite repetition. Where closed time differs crucially from linear time is that the distinction between past and future cannot be maintained. That is the real point of the dwarf's response to Zarathustra. It is consistent with his claim that the apparent linearity of the lanes is a perspectival illusion, like the apparent flatness of the earth's surface as seen by an ordinary observer. At this stage, the perspectival model undermines Zarathustra's polemical strategy and, perhaps for that reason, it is set aside by him from here on.

But is he right to do so? Perhaps it is too soon to concede the point. If the lanes are a perspectival picture, then their properties are determined by the nature of the perspective. Yet we can assert on the basis of experience that no end is to be reached on either lane. As Kant puts it, 'experience teaches me that wherever I may go, I always see a space around me in which I could proceed further'.[28] He is using a similar metaphor to describe the 'space' of possible knowledge. However, whether this space is not only endless but infinite in extent, like the purely logical space of concepts, is another matter. According to Kant's transcendental idealism it is not an infinite plane, but like the surface of a sphere, 'the radius of which can be determined from the curvature of the arc of its surface – that is to say, from the nature of synthetic *a priori* propositions'.[29] For Nietzsche, on the other hand, the perspectival structure of knowledge has an autonomous status. It is not subject to a further schematism in accordance with *a priori* forms of intuition which would replace infinitude with a limited but unbounded form. Thus, the infinite linearity of the two lanes is taken as given in his account.

So too is the difference between past and future, but in a very different way. That is directly encountered in the reality of change, the passage from future to past in the moment. We do not have to look out along the lanes except to find whether either has an end. If the lanes inevitably appear to be infinite, that only says something about the projection. It is always possible to discount a perspectival picture as an illusion, but harder to do so with an experience of force, tension or conflict. Hence, Zarathustra's challenge has not been answered by the concept of circular time.

Zarathustra's Second Challenge

Why was Zarathustra's first challenge to the spirit of gravity so inconclusive? It is not just that an open-ended question allows an evasive answer. A consideration of the lanes alone is inadequate for a recognition of the process of becoming. A more complex vocabulary is needed. What this is shows up as the narrative continues, with a further challenge made by Zarathustra against his opponent:

> 'Behold', I continued, 'this moment! From this gateway, Moment, a long eternal lane runs *backward*: behind us lies an eternity. Must not whatever *can* run have run on this lane before? Must not whatever *can* happen have happened, have been done, have passed by before? And if everything has been there before – what do you think, dwarf, of this moment? Must not this gateway too have been there before? And are not all things knotted together so firmly that this moment draws after it all coming things? Therefore – – itself too?'

This is quite different in style from Zarathustra's first challenge to the dwarf. Instead of a single question, it is a whole sequence of questions. When Zarathustra asked whether the two paths contradict each other eternally, there was no single, clearly correct answer. His challenge called for the exploring of a number of different answers. Here, though, we find questions for which a certain answer is plainly indicated. No time is allowed for reflection, or for successive responses to the questions. Yet the absence of responses is hardly noticed, because they could only be expressions of assent to various assertions made in the guise of questions. We can imagine replies of this kind which, if added to the text, would make the passage run (more or less) along the following lines:

Zarathustra: Behold this moment! From this gateway, Moment, a long eternal lane runs *backward*: behind us lies an eternity. Must not whatever *can* run have run on this lane before?

Dwarf: It seems so, Zarathustra.

Zarathustra:	Must not whatever *can* happen have happened, have been done, have passed by before?
Dwarf:	Very true.
Zarathustra:	And if everything has been there before – what do you think, dwarf, of this moment?
Dwarf:	What do you mean?
Zarathustra:	Must not this gateway too have been there before?
Dwarf:	It must.
Zarathustra:	And are not all things knotted together so firmly that this moment draws after it all coming things?
Dwarf:	True.
Zarathustra:	Therefore – itself too?
Dwarf:	Certainly.

This can be recognized as a Socratic dialogue, along the lines of Plato's. The task set for the respondent is not to answer an open question, but simply to follow an argument. One needs to keep hold of the thread, as Zarathustra put it earlier on, while praising his listeners for *not* doing that. Citing this dismissal of deductive reasoning, T. K. Seung denies that any argument can be present in Zarathustra's second challenge, and asserts that 'His questions are meant to be tentative questions for exploring the plausibility of his ideas concerning the eternal recurrence.'[30] Yet the questions do not look particularly tentative, and they use expressions such as 'If . . .', 'Must not . . .' and 'Therefore…' This is the language of demonstration, not of conjecture. These phrases signal the logical connections between successive propositions, confirming that a line of argument is present in the text.

If there is an argument, what does that imply? The premises are presumably drawn from whatever can be agreed upon by both sides. There is not much risk of error from then on, because each step is secured by a logical inference, and each in turn 'draws after it all coming things'. In other words, the form of the argument corresponds to its own content. In this way, Zarathustra is challenging the dwarf on his own terms, as he is, more generally, by adopting the Socratic style of reasoning. We need not suppose that this is his own preferred mode of thought. In fact, we know that it is not. But given the inconclusive response to his earlier riddle, a change of approach is understandable. His second challenge to the dwarf has no relation to any perspectival standpoint. It is consistently impersonal and universal, as one would expect from a logical argument. His questioning is an *ad hominem* or, in this case, an *ad nanum* strategy. Further, logical necessity has a certain ambiguity. It provides security to the uncertain enquirer, like the thread of Ariadne that shows the way out of a labyrinth, but it can also be used as a means of power over others, compelling them to assent to conclusions that they would otherwise reject. The appeal of rationalism lies in its claim to universal validity. As I noted earlier, the statements of the spirit of gravity in *Thus Spoke Zarathustra* are all universal and impersonal ones.

The dwarf is bound by his own allegiance to such principles to follow the line of thought that Zarathustra presents to its ultimate conclusion.

With such a procedure, there is only one possible conclusion. Hence, it need not be left for a respondent to discover, but can be stated explicitly.

> And this slow spider, which crawls in the moonlight, and this moonlight itself, and you and I in the gateway, whispering together, whispering of eternal things – must not all of us have been here before? And return and run in that other lane, out there, before us, in this long dreadful lane – must we not eternally return?

The dwarf does not answer, and suddenly disappears from the scene. There is, after all, no way of escaping from the force of a logical demonstration, other than by making one's departure.[31] So it is the reader who is left to deal with Zarathustra's challenge.

Before looking more closely at the argument, let us make a general observation. As we have seen, the metaphor of gateway and lanes in 'On the Vision and Riddle' involves distinct approaches, each with its own vocabulary. I call these 'perspectival', 'kinematic' and 'dynamic' models of temporality, since they use the symbolic languages of space, motion and force respectively in representing time. The movement of thought in the central narrative can be characterized in terms of these contrasting models. Zarathustra's first challenge is stated in the vocabulary of perspectivism. For that reason, it is vulnerable to a response which asserts a distinction between reality and illusion. Zarathustra's second challenge cuts off this escape. Without denying the perspectival character of the moment, it puts perspectivism into a further context, making it clear that there is no question of detached observation: we are located in the gateway, but are also among those who will return in the lane 'out there'. The privileged standpoint implied in the dwarf's reply is eliminated by a move from the perspectival to the kinematic and dynamic models of temporality. Only on this second account can the 'contradiction' between the two lanes be appreciated as the predicament of whoever is in the 'moment'. Thus, the first challenge is not an arbitrarily devised riddle. Rather, it presents the quandary that already faces the listener, whether recognized or not. The same could be said of the riddle of the Sphinx, which Oedipus answered successfully. In each case, the listener's understanding of his or her own situation is in question. This is not so different from the task of self-knowledge that philosophy took upon itself both then and later: the uses of language may be different, but the direction of inquiry is much the same.

The Structure of the Argument

There is something familiar about Zarathustra's new challenge, given the discussion of the impossibility of a final state in Chapter 1. The two arguments overlap, and that is not surprising, since they are concerned with the same question about the general form of becoming. The doctrines of creation and a final state, taken together, provide one answer: time has a beginning and end. Having rejected both of these, Nietzsche is left with endless change, but this in turn has two forms, and he sets out to decide between them by using the resources already available. First, the proposition that whatever is possible must already have occurred is asserted, and identified as a consequence of the infinitude of past time. Earlier Zarathustra asserted: 'This long lane stretches back for an eternity. And that long lane out there, that is another eternity.' Now he repeats the same statement: 'From this gateway, Moment, a long eternal lane runs *backward*: behind us lies an eternity.' What follows takes that as given, dismissing out of hand the dwarf's assertion of temporal circularity.

The possibility of a state of rest is not in question here. Instead, the present moment is the subject of this argument, and so the premise needed is simply that this moment is possible. Elsewhere Nietzsche writes, 'Certainly it cannot be left up to the human head to decide whatever is possible and what is not: but in any circumstance the present situation is a possible one, quite apart from our capacity or incapacity to judge concerning the possible – for it is an actual one.'[32] We thus arrive at the conclusion that the present moment must have occurred already. The same can be said of any other state of affairs known to be actual, since those will also be possible by the same principle. But do we know what other states of affairs are actual? Not with the same certainty. In that regard, the present moment has a unique status.

This outcome is not the full doctrine of eternal recurrence, but it already raises issues that have been taken as problematical for the doctrine. Is it possible for events or states of affairs to occur at more than one location in time? Is it possible to *remember* that a present experience has occurred before? I will be addressing each of these in turn.

Any claim that the present state of things has occurred before would be challenging by itself. The thought of eternal return goes further: First, it says that the present state has occurred before infinitely many times, and will do so again infinitely many times. Secondly, it says the same thing about all other actual states. Thirdly, it says that all of these recur eternally in exactly the same order, so that particular sequences of events, such as those that constitute a human life, can be said to recur eternally. The impact of the thought on the individual depends crucially on this last proposition. It would be pointless to have the events of a lifetime rearranged in some different order. It is true that something of the kind has been put forward, originally by Bernd Magnus and later by other writers who envisage random variations in the ordering of the same events.[33] Yet this makes no sense for the case that is of special interest to each of us, the events that together constitute our own lives. Anyone who doubts that should try to rearrange them randomly and consider the outcome. It will become evident that each of our experiences is what it is because of those that have gone before, ruling out any rearrangement.[34]

How do we get from the previous occurrence of the present moment to the eternal recurrence of this and every other moment, in the same order? Nietzsche does not spell out his answer, but the argument he has provided can be extended to supply what is needed. So far, the argument has established that the present moment has occurred before. What is needed is, in effect, an endless repetition of this move. To carry that out, another premise is necessary. It must be assumed that the time between any past occurrence and the present is finite. In that case, assuming again that past time is infinite, the time before any past occurrence must be infinite. In Chapter 1 we encountered the thesis, stated by Aquinas, that even if time is infinite, every past occurrence is located at a finite distance from the present. If Nietzsche accepts this, then the same principle relating possibility and time, that whatever is possible must occur in an infinite past time, can be invoked. We can then conclude that the present moment must have occurred before its past occurrence as already established. Clearly, this argument can be repeated again and again, and that justifies the dramatic move to a further conclusion: that this moment has occurred infinitely many times before in the past.

Zarathustra now adds an equally important corollary, that this moment will occur infinitely many times in the future. Where does this prediction come from? Zarathustra says: 'And are not all things knotted together so firmly that this moment draws after it all that is to come? Therefore – itself too? For whatever can walk – in this long lane out there, too, it must walk once more.' This is a puzzling passage. The last sentence seems to assert that whatever is possible must occur in an infinite future time. That suggests an argument similar to the one already discussed, with a simple substitution of 'future' for 'past'. Yet before this statement there is a cryptic description of things (that is, events or states of affairs) as being 'knotted together'. What is more, this claim is presented as leading to the conclusion that the present moment must recur. How does it do that, though? To sort this out, one needs to realize that the final sentence really belongs earlier in a logical order, as the word 'for' indicates. The idea is not that the present moment will recur because all things are 'knotted together' – by itself, an arbitrary assertion – but rather that, *given that* the moment will recur, it will do so by being 'knotted together' with everything that occurs between its present occurrence. Why is this worth saying? Is not it enough that the moment will occur again? Why account for its recurrence other than through the argument in terms of possibility and occurrence? I think that the answer has to do with the final step that turns the recurrence of one moment into the complete doctrine of recurrence: that is, a recurrence of events in the same order and sequence. When the moment recurs, it will be an outcome of the same preceding history. So it must be not just single events or experiences, but entire human lives, that are the content of the thought of eternal return.

The first two premises in this argument, that is, the infinitude of past time and the principle that whatever is possible must occur in an infinite time is infinite, were considered in Chapter 1, where they figured in the case against the possibility of any state of rest. Within 'On the Vision and Riddle', the infinitude of time has already been asserted in Zarathustra's insistence on the endlessness of the two lanes. I have taken these to be the key to Zarathustra's second challenge to the dwarf. In contrast, Laurence Lampert suggests that Zarathustra is not referring back to this claim, but instead starting from the dwarf's response. He writes, 'The argument presented by Zarathustra intensifies the Dwarf's own thought that time is a circle. From the gateway of the present moment an eternal lane runs backward; if time is a circle, an eternity lies behind us.'[35] Is this a more plausible reading? Lampert is right in identifying Zarathustra's questions as a line of argument, and also right that Zarathustra is challenging the dwarf on his own terms by using argumentation. That does not mean that he accepts anything the dwarf has already said. The weakest point in Lampert's interpretation is his claim that 'if time is a circle, an eternity lies behind us'. This looks like a *non sequitur.* After all, Zarathustra has identified eternity with the long lanes that the dwarf treats contemptuously by asserting that 'time itself is a circle'. Logical reasoning may not be Zarathustra's preferred mode of thinking, but the thoughts here are his own, as he says: 'I was afraid of my own thoughts and the thoughts behind my thoughts.' Hence, I would rather rely on Zarathustra's opening remarks about the lanes, and on the arguments that Nietzsche presents elsewhere, to back up the claim that the present moment has occurred before.

Events and Recurrence

Zarathustra's line of thought has reached its conclusion, that every event or state of affairs must recur endlessly, including the present one in all its detail. Just as it has occurred infinitely often in the past, so it will recur endlessly in the future. This is the thought of eternal return. The first question must be: how is it different from the dwarf's answer? Defending the coherence of circular time is one thing, and identifying it with an eternal recurrence is another. It has seemed natural to many to present this thought as a claim that time is circular in form, or else that it is a cyclical process. The word 'eternal' suggests infinitude, and the concept of a recurring cycle is that of a sequence of events which occur not just once but many times – in this case, infinitely many times. Yet a circle is a closed line which has a finite magnitude, in that two points on it cannot be separated by more than a certain distance. If that is what symbolizes time, then time is finite. Hence, the dilemma seems inescapable: we can speak of a circular time, within which any event occurs just once, or at most a finite number of times, or else of an infinite time within which a finite number of events occur again and again, but not both. In other words, any talk of 'cyclical time' is a sign of confusion. The conclusion of this analysis is that eternal recurrence presupposes an infinite linear time, and that the figure of a circle can serve only to represent the sequence of events that recur infinitely many times. We imagine the circumference of the circle as traversed again and again by the moving finger that points to the present moment. In that case, time is something outside the model, manifested in the added movement of the finger. Alternatively, the model is tacitly expanded to include the finger. On this view, eternal recurrence presupposes a linear model of time. The alternative is a genuinely circular time: that is, a time which is finite, although without beginning or end. In that case there may be a recurrence of the same states of affairs, but it will not be eternal recurrence, since they will occur only a certain number of times.

A common thesis in readings of Nietzsche is that any argument for eternal recurrence is, in fact, an argument for something else – namely, for circular time. Crucial to this view is the claim that any recurrence of the same event is simply impossible. The move has an air of arbitrariness about it, but there is more to be said. Twentieth-century analytic philosophers pursued a research programme which aimed at conceptualizing events as *basic particulars,* along the same lines as material objects.[36] W. V. O. Quine described this project as a 'reification' of events, and praised it as bringing a wider area of language within the scope of standard predicate logic – for instance, by replacing adverbs with corresponding (but semantically more manageable) adjectives.[37] One part of this approach is the thesis that an event can occur at only one time, just as a material object can be located in only one place. It is astonishing that otherwise acute Nietzsche scholars have accepted this proposition without any question.[38] For it is not an innocent piece of conceptual analysis, as they seem to suppose, but a programmatic thesis involving a move away from ordinary ways of speaking and thinking about events.

Common usage suggests that an event *can* recur, that is, occur more than once. Those wanting to interpret events as basic particulars are prepared to give this up in the interests of what they believe will be a more systematic and consistent ontology. Not all analytic philosophers have been prepared to make such a sacrifice. Roderick Chisholm introduces a discussion of events by saying: 'Any theory of events should be adequate to the fact of *recurrence,* to the fact that there are some things that recur, or happen more than once.'[39] Is it a fact, though? For advocates of a reification of events, our everyday use of this word is a loose one. What we refer to as a 'recurrence' is not another occurrence of the same event, but only the occurrence of a similar event, or an event of the same kind. Either way, the claim being made about 'recurrence' is not that it is ambiguous, but that it has no application in its literal sense. So it should be admitted that to say that events cannot recur is to make an unusual claim, one that needs justifying. This is not to say that ordinary language is never in need of correction, or has an authority that overrules philosophical arguments, only that some reason is normally needed for subjecting it to major change.

Is there such a reason where recurrence is concerned? In an often cited analysis, Bernd Magnus argues that the principle of the identity of indiscernibles provides a demonstration of the impossibility of recurrence.[40] The principle (in its converse form) asserts that the same things must have the same properties. Where events are concerned, one property is that of occurring at a certain time. Presumably, then, we can say that the same event must occur at the same time. Now, does this imply that an event cannot occur at several different times? Not at all. An event which occurs at one time can be identified with an event which occurs at another time, provided that the event which occurs at the first time also occurs at the second time, and the event which occurs at the second time also occurs at the first time. This is not to say that the times are the same, as Robert Gooding-Williams suggests when he writes, 'When an event recurs, it must recur at the same time it originally occurred, for otherwise it would not be the same event.'[41]

Now, it is not surprising that a philosophical programme aiming at a 'reification' of events should propose a criterion of identity for events which is like that for objects. Identity of location in time and space is a case in point; but as we have seen, the proposition that the same event must occur at the same time does not rule out recurrence, unless it is arbitrarily supposed that 'time' must refer to a single location. One needs to ask whether a deliberate reification of events provides an appropriate approach to Nietzsche's ideas about time, events and recurrence. Nietzsche has no objection to conceptual revision. He wants to change, even get rid of, many everyday concepts. But the direction of his project is just the opposite of an extension of the 'atomistic' model of basic particulars. He is critical of that model on its own ground, and tries to undermine its credibility by charging that it rests on unjustified assumptions about the 'self' as a paradigm case of substantiality. Quine's assumption that we already have a sound ontology whose scope can be usefully broadened by a semantics of events would strike him as a false trail.

What Nietzsche wants is not to consider events as things but to consider things as processes. This is a far more problematical proposal, given that our present conceptual vocabulary is almost entirely formulated in terms of things and their actions on one another. Nietzsche likes to blame the Indo-European languages for our situation, and seems to imply that a language like that of Jorge Luis Borges' Tlön (in which the equivalent of 'The moon rose above the river' would translate literally as 'Upward behind the onstreaming it mooned'[42]) might avoid the problem. But he locates the real issue at a deeper level of motivation, in cognitive traits shared by other sentient species whose behaviour is guided by recognition of recurring features of their environment, a trait selected for its usefulness, not its claim to objective truth. Our grammar is only a symptom of this problem.

A consideration of events and processes is helpful in addressing Nietzsche's ideas on time and recurrence. To the question: *what* is it that is supposed to recur? he gives varying answers. The most common are the obvious choices, 'everything'[43] and 'the same'.[44] Next come a family of related expressions: 'state'

(*Lage*), 'state of force' and 'total state',[45] 'state of affairs' (*Zustand*)[46] and the more specific 'combination'.[47] This group can be taken to include occasionally used terms such as 'case' (*Fall*),[48] 'form' (*Form* or *Gestalt*),[49] 'product',[50] and other similar words. Finally, there are references to the recurrence of particular things such as sun, earth, eagle and snake, spider and moonlight,[51] as well as experiences such as pleasures and pains, thoughts and emotions.[52]

There is a certain association between these groupings and the words that Nietzsche uses for 'recurrence': *Wiederkehr* (return), *Wiederkunft* (recurrence) and *Wiederholung* (repetition). Joan Stambaugh says that 'Nietzsche almost never used the word "repetition" (*Wiederholung*).'[53] This is true of his published works, although it does appear in notebook entries. Some believe that these expressions, or at least the first two, have significantly different meanings.[54] Their correlations with the other terms are relevant to this issue: Nietzsche uses *Wiederholung* for states and similar relatively abstract and impersonal concepts, but *Wiederkunft* for the more personal examples of the last group just mentioned. Apart from broad references to 'everything' and 'the same', he uses *Wiederkehr* in a few places only, notably for the 'moment'.[55] Yet he is not systematic in all this, and it is possible that such associations have no particular importance. Hence, a theory based on them needs to be treated with caution. Promising though it may seem, an all-purpose semantics of events is not going to solve our problems with Nietzsche's thought of eternal return. In any case, as Chisholm's declaration makes clear, the presumption that such a semantics will rule out recurrence in a literal sense is a hasty one.

Chapter 7

The Eternal Perspective

What does the doctrine of eternal recurrence mean for our relation to past, present and future? The importance that Nietzsche places on the concept makes the question harder to answer, because it leads him to present it in ways that emphasize its potential for personal transformation above everything else. This chapter, however, will address epistemic questions before turning to aspects of the doctrine's significance for human life.

One issue already mentioned is that of memory. Is it possible to *know* that a present state of things is the recurrence of some previous state, the same in every detail? Memory would be direct support for recurrence, not subject to the uncertainties of the complex and incomplete arguments found in Nietzsche's writing. It would be like the evidence cited for reincarnation when people claim to remember episodes from past lives. In those cases it is possible, in principle at any rate, for a degree of confirmation to take place. If I remember having been Napoleon, for example, I might make memory statements whose accuracy could be verified, putting sceptics on the defensive, at least until another person with an equally good claim turns up. For a past life exactly similar in content, however, this would be a meaningless exercise. How can anyone tell whether what I am remembering is a past episode in my present life or in its previous counterpart? The interesting claims would be those that are now predictions, that is, which have not yet occurred in the present cycle.

This is just what Paul Loeb envisages. He suggests that Nietzsche 'thinks each one of us holds a memory of our life's eternal recurrence'.[1] This memory manifests itself from time to time as intimations of parts of our life that are still to come and thus, on the usual view, unknown. Hence, they appear as uncanny experiences of foresight. The historian may be a prophet looking backwards (in Friedrich von Schlegel's words) but so is the prophet himself, in this case. Loeb asserts that without memory or recognition, or more generally direct *awareness*, covering future experiences as well as past ones, the doctrine of eternal recurrence is 'insupportable, insignificant, and incoherent'.[2] If he is right, then the central part of *Thus Spoke Zarathustra* stands or falls with the question of memory. I think this claim needs further support, but it is not necessary to raise the stakes so far to find the issue of memory important, although not necessarily in just the way that Loeb identifies.

People sometimes do have an impression that the present experience has occurred before, in every detail. One philosopher who discusses this phenomenon (while not having experienced it himself) is Henri Bergson. In an article on 'false recognition', he explains what is meant.

> Some one may be attending to what is going on or taking part in a conversation, when suddenly the conviction will come over him that he has already seen what he is now seeing, heard what he is now hearing, uttered the sentence he is uttering, – that he has already been here in this very place in which he now is, in the same circumstance, feeling, perceiving, thinking and willing the same things, and, in fact, that he is living again, down to the minutest details, some moments of his past life.[3]

The label 'false recognition' begs the question about any claim to truth, but Bergson says that 'we know full well that no life goes twice through the same moment of its history, that time does not remount its course'.[4] He explains that this apparent memory is not referred to any particular part time. It is not that the present perception reminds us of something that has happened before – that is a common experience, without any paradoxical character. Rather, the encompassing quality of this recognition means that there is no separate indication of time. Bergson now makes a bold claim that, far from being an unusual and hard-to-explain experience, this is a very common one, or rather would be common if there were not powerful mental forces acting to suppress it nearly all of the time.[5] He proposes that every perception is accompanied by memory from the beginning: it is not that memory is a separate event coming later. However, ordinarily our attention is so much on the perception that we do not notice the accompanying memory. Only when the primary focus is weakened for some reason (such as tiredness or distraction) does the memory make an appearance in consciousness alongside and simultaneous with the perception. The result is the experience of 'false recognition'.

With a psychological explanation available, whether Bergson's or one of the other theories that he mentions, why should we take 'memory of the present' as evidence for anything as contrary to our conception of reality as eternal recurrence? As for anticipations of experiences that have not yet occurred, supposing that these turn out to be reliable, how can they be explained? A naturalistic account of prevision (that is, a direct intuition of future events, as distinct from a prediction based on evidence) seems unlikely but, again, may well look more plausible than eternal recurrence. In Loeb's view, the *daimon* in section 341 of *The Gay Science* is bringing about a recollection of the previous occurrence of that moment and, in turn, of the whole life of which it is a part. How does the *daimon* do this? Not by the Socratic or Freudian method of subtle questioning that leads a respondent to bring forth hidden knowledge of his or her own accord, but by a sudden and violent confrontation. The status quo of personality has to be disrupted for the true self to emerge. And yet it is not indicated in *The Gay Science* that what the *daimon* produces in the listener is memory, although he assumes the hearer's knowledge of his or her present life, 'as you have lived it'. Neither of the responses that Nietzsche describes incorporates a recollection: they are positive or negative reactions that depend on having beliefs, but not on having knowledge. (Use of the word 'awareness' glosses over this distinction.) The response 'You are a god, and never have I heard anything more godlike' would be excessive praise for a mere prompting of memory. As for believing that what the *daimon* says is true, we often accept things on the testimony of others, even about events that we might remember but do not. To see a photograph from one's childhood, for example, is not to recall the occasion on which it was taken. On the other hand, it might revive a real memory. So too, none of this rules out the possibility that, in consequence of the *daimon*'s announcement, I *do* now remember past occurrences and foresee future recurrences of the present moment. Given that scenario, the idea of memory of a previous occurrence needs to be considered further.

Pierre Klossowski also defends memory in relation to recurrence, but in an unexpected way. He argues that we need to remember *in order to forget.* Like Loeb, he claims that insight into the eternal return is *anamnesis* – that is, remembering that this present event has happened before. The word suggests Plato's account of knowledge, but Klossowski's idea is different. His real interest is in the importance of forgetting. He asserts that in realizing the thought, we remember, but only in order to forget again straight away, for 'how could the return not bring back forgetfulness?'[6] Klossowski also thinks that the thought of eternal return accomplishes an abolition of personal identity.

> At the moment the Eternal return is revealed to me, I cease to be myself *hic et nunc* and am susceptible to becoming innumerable others, knowing that I shall forget this revelation once I am outside the memory of myself; this forgetting forms the object of my present willing; for such a forgetting would amount to a memory outside my own limits; and my present consciousness will be established only in the forgetting of my other possible identities.[7]

Klossowski's analysis is driven by his insistence on a renunciation of personal identity, which depends on the guardianship of memory and is therefore directly negated by a will to forget, or rather to transpose any memory beyond the limits of one's present identity. Discussing related issues, Stanley Rosen writes, 'The puzzle for the reader of Nietzsche's work is how one can both recollect and forget the doctrine of the eternal return.'[8] His solution is that 'the one who wills or remembers is distinct from the one who, having forgotten, creates'. That fortunate error is confined to a select group, likened by Rosen to Plato's privileged class of philosophical adepts. Only by virtue of their ignorance of recurrence (understood here as a fact about the universe as a whole) are these 'supermen' capable of transcending nihilism through creative action, and so initiating a new historical period.

As one can see, there is no agreement on why the issue of memory is important, but it must have to do with individual personality, life and activity. The question is, why *memory*, and not just a generic awareness? In Loeb's argument, the reason seems to be that memory makes a greater impact on the individual than awareness or even belief, taken by themselves. That is presumably because of the element of *givenness* that memory involves, giving it a certain parallel to perception. Whatever the reasons for its importance, the possibility of memory raises important issues about eternal recurrence as a version of temporality, and so I turn now to these.

The Case against Memory

If memory is impossible, then the question of its consequences does not arise. Among writers on Nietzsche, any notion of remembering past cycles of recurrence is commonly rejected out of hand. The rationale is that if the *same* state of affairs recurs again and again, it must be impossible to remember its past occurrences, or to foresee its future occurrences. That is, a recurrence of the *same* events or states of affairs rules out any cognitive relation between one and another. One of the earliest to argue this case was Georg Simmel, who wrote:

> From a logical viewpoint there is no inherent importance to repetition, because no synthesis is possible of successive repetitions. If an experience is repeated within my existence, this repetition can be of enormous importance for me, but only because I remember the first instance and only if I have already been altered by it. But if we assume the empirically impossible case in which the second instance finds me in exactly the same state as I was when the first instance occurred, then my reaction the second time would be the same as it was initially and, therefore, the repetition would have no importance for me.[9]

Conversely, Simmel argues, any experience which does point back to an earlier one must be different from the previous experience 'just by virtue of that acknowledgement'. Recognition adds something new by attaching a particular meaning to the content that has reappeared. Hence, there is no recurrence of the experience in exactly the same form.

This has seemed to many readers (including myself) to be a conclusive argument. However, Loeb draws on a response given by Philip J. Kain to rebut the case against memory of recurrence.[10] Kain argues that it presupposes an *initial* occurrence that includes no memory, so that this is an additional element in subsequent recognition of the experience. If however we suppose that the experience, *including* memory, has always occurred throughout an infinite past, then there is no difference between successive instances after all. This argument appears to be a strong response to the common view, and so we need to address the issue more closely.

What is Nietzsche's own opinion about memory and recurrence? Where he does touch on this subject, what he says does not support claims for a memory of past occurrences of the present moment. For example, he draws attention to the incompatibility of memory and repetition in an 1881 note: 'The inorganic world, even if it was mostly organic, *has learned nothing*, is always without a past! If it were otherwise, there could never be a repetition (*Wiederholung*) – for there would always arise from matter something with *new* qualities and *new* pasts.'[11] The distinguishing feature of the organic, according to this text, is that it learns and 'has a past'. In this sense, every organic being can be said to have memory, even a kind of 'mind' (*Geist*).[12] This is why it does not repeat itself exactly, but rather 'becomes'.[13] That is not to say that it has conscious memory. Even for sentient beings, recognition of something does not imply a recollection of the experiences from which its familiarity is derived. On the contrary, habits and conditioned responses allow us a high degree of forgetfulness in relation to their sources, and this economical policy is a practical advantage, even if it leads to misconceptions about the nature and origin of many human concepts, especially the moral sentiments.[14]

Nietzsche is also a sharp critic of memory where it involves judgements of sameness. Recognition is essential to learning and thus to survival, but it depends on limitations of perception which are also the conditions of a form of life. He insists that 'The similar is not a degree of the same: rather something totally different from the same'[15] and argues that we are usually mistaken when we take things to be the same: 'in fact nothing is the same as anything else'.[16] He mounts an even more striking argument against sameness by proposing a strong principle: that nothing can be the same unless *everything* is the same.[17] The outcome is a strange one: we have a concept of identity, but can never apply that category to our experience. Nietzsche's critique of identity gives rise to a puzzle over eternal recurrence. How can he, on the one hand, claim that no two things are ever the same and, on the other, that everything recurs in just the same form, down to the smallest detail? If nothing in experience is ever the same as anything else in experience, it must follow that recurrence is *not* experienced. It may be *thought*, but that is another matter, or so it seems.

In considering this issue, we should take account of another feature of Nietzsche's presentations. The doctrine is always expressed in *modal* language. Zarathustra does not say that the gateway *has* been there before; he says that it *must have* been there.[18] Similarly, he does not say that the moment will return, but that it must return. Again, the *daimon* announces that 'this life as you now live it and have lived it, you must live once more and innumerable times more'.[19] What the doctrine asserts is not just that my life will recur in the same way, but that it must recur in the same way. It presents recurrence not as a fact but as a necessity.

How is this related to our ability to remember a past occurrence or foresee a future recurrence of this moment? We do not say that something 'must have happened' when we are remembering that it did happen. One of Nietzsche's best-known aphorisms displays this contrast. '"I have done that", says my memory. "I cannot have done that," says my pride, and remains inexorable. Eventually – memory yields.'[20] Whereas memory says 'I have done that', pride uses a different kind of wording, 'I *cannot* have done that.' This response is stronger than a simple negation. The force of necessity is expressed in its 'cannot', and it is this force that enables pride to be 'inexorable'. Further, the 'cannot' implies a reason for its repudiation of memory. We can guess what the reason is: 'I am too fine a person to do such a thing.' When memory yields in the face of this demand, the further outcome is a lie, first to oneself and then to others: 'I have not done that.' This assertion conceals its origin by adopting the form that is characteristic of memory, thus acting as a substitute for 'I have done that'. What we remember has the same factual character as what we experience. Memory tells us what has occurred, just as perception tells us what is occurring. It is not an inference from a present experience, such as a particular kind of mental image, one that bears the mark of pastness. As philosophers such as Bertrand Russell have pointed out, for any such image to be evidence of the past, there must be a way of establishing the relation between the two, and that would require access to past objects apart from the mental image.[21]

This direct relation to the past may seem a point in favour of memory as validation, but it is also the opposite. Memory is as fallible as perception, and Nietzsche considers that perception is very fallible indeed. So, when pride overrules memory, could it be that pride is in the right, rather than being (as a first reading of the aphorism suggests) a bad influence? Similarly, could an inability to recall past occurrences of present experiences be a liberating factor, enabling us to hear the message of the *daimon* with an open mind? There is still a case against memory here, even if the most common argument is mistaken.

If memory has consequences, so does lack of memory. The absence of any memory of a recurrence rules out the idea of *counting* cycles of recurrence. If one assumes that the process has a starting point, then there will be a first cycle, a second, and so on. But as Kain asks, what if the sequence has no beginning and is postulated as having existed throughout a past eternity? How can we tell the cycles apart? The reply may be that we simply start at a given occurrence and then proceed to count the number of times around the circle of becoming. Linda L. Williams and Joseph T. Palencik use this idea to rebut an identification of eternal return with circular time. After claiming that 'Nietzsche tries to prove that time is circular', they argue in opposition to this model that 'the fact that we can count the number of times around the ring is sufficient to differentiate the occurrences, which means that they are not absolutely identical'.[22] But is it really a fact, or does it assume an external standpoint whose existence is highly doubtful? After all, if we are located within the circle of recurrence, then our mental acts cannot have a different content in successive cycles. As Alistair Moles points out, if I count 'one' now, I will not be counting 'two' when the same total state comes round again.[23] I will be counting 'one' again. I cannot even count 'one', and genuinely mean it, if I am simultaneously recalling a previous occurrence of the present state of things, since I know better: I am aware that this is not the first time for anything, including counting.

Even so, cannot we imagine an external standpoint for the sake of argument, while conceding that it is not possible to occupy such a position? The problem now is that the imagined observer is supposed to be taking a particular cycle as the starting point for counting subsequent ones. But how can the observer be referring to one cycle rather than another? This standpoint is a timeless one, standing apart from the course of becoming. Hence, its occupant cannot be fixing on the *present* state of affairs as the starting point, because if this being had the same past, present and future, then it would be within the cycle, not outside it. There are no causal links to fix the reference of a proper name – and description will not be enough, on the hypothesis of identical recurrence. Compare this with Nietzsche's own observer who counts forward or backward in time and, finding no beginning or end, concludes that time is infinite. That is an external standpoint, but only in a relative sense. It is not on either of the lanes, but that is not because it stands outside the world of appearance: rather, it is because it is a perspectival standpoint, located in the gateway.

Past, Future and Eternal Recurrence

So far I have criticized several approaches to the eternal return. The doctrine's puzzles and difficulties are evident in a literature marked by a wide variety of interpretations, as well as a high incidence of partial readings, arbitrary assumptions and unsound arguments. Having said this, I recognize the need for an analysis that avoids these pitfalls and makes better sense of what Nietzsche actually says. What follows will be an attempt to set the doctrine within the framework of ideas presented in previous chapters or, rather, to develop these ideas further to take it in. Even if there are several ways of grasping the thought of eternal return, I believe that some valid guidelines can be formulated.

First of all, any interpretation that fails to include *becoming* must be mistaken. It may seem unlikely that anyone could adopt such a view, but some writers have invoked an eternalized present raised above any flow of time. Robert Gooding-Williams, for example, takes Zarathustra to be 'teaching himself how to envision his soul as the omnipresent incarnation of an eternal "now"' and 'teaching himself to see his soul as a pantheistic *nunc stans*'.[24] T. K. Seung argues that the thought of eternal return is at best a poetic image or metaphor, and that Zarathustra's vision is of a timeless, eternal domain in which nothing can happen or become, let alone return. 'Because all things are eternally present in the eternal ring,' he writes, 'there can be no repetition.'[25] My approach is very different from these. However eternal recurrence is understood, it must be an affirmation of becoming.

Secondly, it is essential to recognize that the doctrine is about past, present and future. Here again I differ from other writers. Most commentaries are dominated by a model of time using the vocabulary of McTaggart's B-series – that is, which places events at earlier and later times in a single sequence. Debate then centres on the possible identity of events located at different positions in this sequence, or else on the coherence of thinking of the sequence as circular in form. These issues make sense in their own terms, but Nietzsche's ideas on temporality are presented in a different conceptual and symbolic language. This is evident in the metaphor of the gateway 'Moment' and lanes standing for past and future time, and in the ominous prediction of the *daimon* of *The Gay Science*, that 'This life as you now live it and have lived it, you will have to live once more and innumerable times more.' McTaggart's A-series of past, present and future is Nietzsche's standard way of stating the doctrine of eternal recurrence. Is this true of the places that contain arguments in its support? Well, it fits the passage in 'On the Vision and Riddle' that we have examined, and that is a line of thought, not simply an announcement. As for related notebook entries, one can reply that many of these are concerned with general presuppositions of the doctrine, rather than the doctrine itself. Not surprisingly, they are concerned with broad conceptions of time, space and force. Hence, they are not where to look if one wants to find the specificity of the doctrine.

Another way of stating my second thesis is to say that the thought of eternal return is primarily expressed in *tensed language*. Confirmation is to be found in every one of the presentations that Nietzsche published, starting with those just cited. Again, it is only the drafted arguments giving theoretical background that lack this feature. Corresponding to this is the unmistakably tensed formulation of Zarathustra's concept of redemption from the spirit of revenge: 'All "It was" is a fragment, a riddle, a dreadful accident – until the creative will says to it, "But thus I willed it." Until the creative will says to it, "But thus I will it; thus shall I will it."'[26] The parallel with the emphasis on tense in the doctrine of eternal return highlights the bond between them. One defines the question and the other provides an answer, although just how it does this remains to be seen.

The tensed formulation of eternal recurrence tells us a lot about its implications. Most of all, it points to the importance of the doctrine for human life. Tensed language has an essential connection with our hopes, fears and regrets in a way that talk of 'before' and 'after' does not. We do not care about whether one thing comes before or after another in the way that we care about whether something is in the past or future. In the words of A. N. Prior, 'Thank goodness that's over!' does not mean 'Thank goodness the conclusion of that thing is contemporaneous with this utterance.' As he comments, 'Why should anyone be thankful for that?'[27] The impact of the thought of eternal return depends on this aspect of its formulation. On a tensed interpretation, recurrences of the present experience are to be taken not as occurring earlier or later than this moment, but as occurring in the past and future – or rather, as having occurred in one's past and still to occur in one's future. It is always an endless prospect ahead of the hearer. This is how the eternal return is presented in *The Gay Science* and *Thus Spoke Zarathustra*. What is said is that we will relive every experience that we have had, and the 'again and again' is a repetition of the future tense rather than of the 'later than' relation, with its associated extensive magnitude, an intervening period of time. It is, so to speak, a raising of 'futureness' to a higher power, rather than just an addition or multiplication. Similarly with pastness, where a reiteration is already expressed in our language by the pluperfect tense. That the focus of attention in these texts is the doctrine's impact supports the significance of tensed formulations. On my interpretation, the eternal return does not mean that any state of affairs occurs before and after its current occurrence, but rather that it has occurred in the past and will occur in the future, or, better still, that it must have occurred in the past and must be still to occur in the future.

This insight enables us to see what is wrong with the interpretations that first identify the eternal return with circular time, and then struggle to explain its significance for human life. They have left out the perspectival standpoint that in turn arises from the interplay of forces within and acting on the individual. The effect of the thought on those who encounter it may seem a separate subject, a psychological rather than philosophical question, but on reflection it is another aspect of this same one. The issue is what it says about time, and what that means for human beings. In *The Gay Science* we are presented with several possible responses to the *daimon* who announces the thought. For ourselves the text makes a menacing prediction: 'If this thought gained possession of you, it would change you as you are and perhaps crush you.'[28] We need to look into these alternatives, and the rest of this chapter will undertake that task.

First of all, however, another option that does not appear in this text should be considered. It is that someone who is confronted by the demon's announcement may treat it, in Ivan Soll's words, as 'a matter of complete indifference'.[29] If it is possible to justify this response, then the options presented by the *daimon* can both be dismissed as expressions of conceptual confusion. 'Indifference' means not just casual unconcern, but a more considered attitude, supported by an argument raised in Georg Simmel's 1907 critique of the doctrine of eternal return.[30] A forceful presentation is found in a 1913 article by Henri Bois, who argues that 'In the end it is not me who returns, it is someone else who "resembles me like a brother", but it is not me, because memory is lacking.'[31] Bois signals his intention to respond to Nietzsche's challenge with a polite rejection: 'Pardon me, I will reply, it is not my life that will return, but another life which will resemble it without being it, and which will concern a being similar to me, but who will not be me. In short, it is only my present life in the present world which concerns me.'

The indifference indicated here is not the heroic equanimity of the Stoic tradition, epitomized by Horace's man of virtue and integrity who remains impassive even if the heavens fall upon him.[32] Rather, it is an eliminative strategy that aims to demonstrate the unreality of the threat to the individual person, so that indifference requires no special moral qualities but only rational sense. The classic Epicurean argument about fear of death – 'Where death is, I am not, and where I am, death is not' – is like this. Its conclusion is that indifference is the only sensible attitude towards the prospect of death, since the necessary conditions for caring about the further course of events – for taking them personally, so to speak – are lacking. The same claim is made about the eternal return by a line of interpreters, starting with Simmel and Bois, and extending to later writers such as Soll and Maudemarie Clark.[33] They argue that we should be indifferent to the prospect of an infinite recurrence of the present state of things, since we cannot identify ourselves with the person who figures in these future states, however exactly similar in every detail he or she may be.

If this recommendation is accepted, what will its consequence be? That person will go on just as before. Nietzsche never mentions this possibility, but I think his account can be widened to take it in, even if others have taken the point as a telling objection. He says that the doctrine is 'mild' towards those who do not believe it, since it makes no threats of future punishment, but leaves them with their 'fleeting life'.[34] These people may have long and contented existences, leaving behind many offspring who follow the same way of life. Zarathustra's description of the complacent and prolific 'last men' makes this plain. So, what would Nietzsche say about Bois' claim that 'It is only my present life in the present world which concerns me'? Probably that it is an attempt to set up what is close at hand, the temporary well-being of individuals and social groups, as the sole standard of value. A secular attitude of this kind turns its back not only on traditional belief in a higher realm but also on the thought of eternal recurrence.

To understand our feelings about recurrence on the model of fear or hope about some experience that we expect to have at a later time is to open up the possibility of indifference. Here recurrence is defined as an eventual actuality with which we may or may not be able to identify ourselves, following some argumentation on the issue of personal identity. A different view arises if we take the tensed interpretation of the doctrine seriously, rather than relying on relations of 'before' and 'after'. In that case, recurrence is a future prospect from the moment in which we learn of it, and must be confronted by the individual person in one way or another.

The End of Time

Nietzsche clearly thinks that the thought will have a dramatic effect if it is understood and 'incorporated'. One 1881 notebook entry, however, seems to endorse a startling neutrality of content as his recommendation for approaching the conduct of life in the light of the eternal return:

> My doctrine declares: the task is to live in such a way that you must wish to live again – you will anyway! He to whom striving gives the highest feeling, let him strive; he to whom rest gives the highest feeling, let him rest; he to whom organisation, allegiance, obedience gives the highest feeling, let him obey. Only *let* him become aware of *what* gives him the highest feeling and avoid no means! Eternity is at stake![35]

Readers have been puzzled by the first sentence, which sets a 'task' for listeners to fulfil, and in the next breath tells them that their future is already set in place. It is hard not to see these as inconsistent statements. What follows is equally striking. Joan Stambaugh writes, 'At first reading the gist of this fragment seems to be: it doesn't matter what you do, whatever makes you feel good, go ahead and do it!'[36] As she remarks, this suggests a rather shallow and not particularly interesting piece of thinking on Nietzsche's part. We need to note, however, that the text echoes the final chapter of the Book of Revelation: 'He that is unjust, let him be unjust still: and he that is filthy, let him be filthy still: and he that is righteous, let him be righteous still: and he that is holy, let him be holy still.'[37] This message comes from the angel who proclaims the completion of human history. It is spoken when 'the time is at hand', that is, when the Last Judgement is imminent. What it says is that there is no longer any point in reforming one's life, given 'that there should be time no longer'.[38] The accounts are closed and not subject to further addition or subtraction, for time as we have known it (that is, the present age) is at an end. Instead of an open future, all that lies ahead is God's judgement on humanity, followed by reward for the righteous and holy, and punishment for the unjust and filthy.

Is that what happens in the affirmation of eternal recurrence? Nietzsche's simulation of a scriptural text has a purpose, even though he is setting out a radically different alternative. His understanding of the doctrine as an *event* is closely linked with the Christian conception of an end of this world, which coincides with a dramatic personal transformation: we do not cease to exist, but we become something else. For Nietzsche, this is what happens when the thought of eternal recurrence is incorporated and allowed to exercise its power. The eternal return is not a final state of rest, and yet in some sense it can be said to indicate the end of time. To understand his adaptation of the Biblical passage, we need to appreciate that he is attempting an alternative to two philosophies of life and their corresponding models of temporality. The note begins:

> The political madness which I smile at, just as contemporaries smile at the religious madness of earlier times, is above all secularisation, faith in the world and an exclusion of the 'beyond' and the 'backworld'. Its goal is the well-being of the *fleeting* individual; thus socialism is its fruit, i.e. fleeting

> individuals want to win their happiness, through socialisation, they have no reason to wait like people with eternal souls and eternal becoming and future improvement.[39]

Nietzsche compares himself with his contemporaries who, he says, smile at the religious faith of past times. In *The Gay Science* this dismissal of religious belief is exemplified by the crowd in the marketplace (including 'many of those who did not believe in God') who are incapable of taking in the message of the death of God, even though that event is their own doing.[40] However, what he calls 'religious madness' has more to do with personal immortality and the hope of eternal happiness or threat of eternal punishment.

Zarathustra comforts a dying man by telling him, 'Your soul will be dead even before your body: fear nothing further.'[41] 'Backworld' (*Hinterwelt*) is Nietzsche's derogatory label for the concept of a reality beyond experience, including a future life – in fact, one of Zarathustra's early discourses is aimed against its defenders.[42] So, why does Nietzsche object to the alternative presented by secular humanism? As a philosopher who belongs to 'tomorrow and the day after tomorrow', he is just as dissatisfied with this modern orthodoxy. From his standpoint, a dedication to 'the well-being of the *fleeting* individual' is as misguided as following a promise of eternal happiness. He calls the pursuit of well-being 'madness' because there is something paradoxical about taking the individual's quality of life as a guiding ideal. A few pages later he returns to the same theme and brings out this irony: 'Religious faith declines, man learns to regard himself as fleeting and as inessential, so in the end he becomes weak; instead of practicing striving and enduring, he wants present enjoyment, he makes things easy for himself – and maybe he uses a lot of spirit in the process.'[43] Nietzsche blames the socialist movement for encouraging this tendency, on the rather 'Naumburgian' assumption that the aim of socialism is a life of ease for the common masses.

What is so bad about happiness in this life? one may say. Nietzsche's answer depends on an unexpectedly balanced view of religion's contributions to human life. In later writings, he argues that the will to truth, a drive that is hard to explain as an evolutionary adaptation, owes a great deal to religious traditions, even if it has ended by turning against their doctrinal content. Usually, though, the emphasis is more on aesthetic experience. In *Human, All-Too-Human* he argues that religious beliefs, however false, have inspired many of the greatest works of art.[44] In *The Gay Science* a close parallel is drawn between art and religion, and this time it is art (and theatre in particular) that is taken as primary. Art gives us the ability to see ourselves from a distance, in a form that simplifies and transfigures, giving access to traits that would otherwise go unobserved, given our usual preoccupation with the immediate concerns of life. Nietzsche now makes a comparison with religion:

> Perhaps one should concede a similar merit to the religion that made men see the sinfulness of every single individual through a magnifying glass,

> turning the sinner into a great, immortal criminal. By surrounding him with eternal perspectives, it taught man to see himself from a distance and as something past and whole.[45]

This appreciation is the key to Nietzsche's rejection of secularism. It is not just that it tends towards 'socialism', but that it narrows one's perspective to what is at hand, in both space and time. The details of his critique are found in his discussions of utilitarianism and its psychology of pleasure and pain.[46] Here he is thinking of the bigger picture, and the need for a longer perspective that acts as a magnifying glass or a telescope. The eternal return is intended to do just this as well, and that is why the parallel with the angel's message is important. Despite its rejection of personal immortality and eternal reward or punishment, the thought of eternal return has something in common with the religious attitude: both take the 'eternal perspective'. Our everyday perspective is narrow, partial and fragmentary, restricted to a here-and-now in which everything is incomplete and still becoming. The eternal perspective enables us to get a complete picture. It allows us to see our lives from a distance and as 'something past and whole'.

Why should this be such a shattering experience? The answer is linked with another theme prominent in Nietzsche's thought. *Fatalism* comes in several forms: it may be based on divine foreknowledge, or causal determinism, or reasoning about possibility, past and future. But each of these entails a determination of the future by the past. Hence, if the past is fixed and necessary (an assumption, but a very plausible one) the same thing must be true of the future as well, given this relation. Now, the doctrine that whatever happens has already happened before, and must of necessity occur again in the future, also confers a privileged status on the past in relation to the future. This is hinted at in the pessimistic pronouncement that 'All is empty, all is the same, all has been.'[47] This predominance of pastness is what produces an unsettling effect, in the same way as other kinds of fatalism. It is the key to the claim that one may find the message unbearable, a weight too great to be borne. In Nietzsche's earliest notes on the eternal return, the theme of fatalism is a main concern, and the doctrine is introduced as fatalism in its 'most extreme form'.[48] For it implies that our own deliberation and decisions for the future have already been made during those past cycles, and will inevitably be made again.

Once the doctrine acquires its own significance, over and above being a version of fatalism, great importance is placed on its possible consequences. Nietzsche claims that many could perish in consequence of the thought of eternal return: not just individuals, but even whole societies, and that for this reason the doctrine's appearance is a turning point in human history.[49] These forecasts are largely based on his reading of W. E. H. Lecky's account of the Methodist reformers of the eighteenth century, who reduced large congregations to a state of emotional frenzy and physical collapse when they preached on eternal punishment.[50] Having presented Zarathustra as a conspicuous failure in addressing a public audience, Nietzsche could hardly credit him with this kind of success later in the work. In any case, he had repudiated popular appeal as a criterion of achievement in expressing insights. The only person who undergoes such a collapse is Zarathustra himself, after an intense but inconclusive confrontation with his 'abysmal thought'.

Still, the parallel with eternal punishment is important. The claim that 'eternity is at stake' (*Es gilt die Ewigkeit*) applies to one's response to the eternal return, as it does with religious belief. Eternity figures prominently in Nietzsche's earliest notes on the doctrine. He writes, 'Let us impress the image of eternity on our life! This thought contains more than all religions that despise this life as something fleeting and teach us to look to an indefinite *other* life.'[51] The calculated use of the Platonic expression 'image of eternity' supports Stanley Rosen's statement that Nietzsche's vision 'retains a Platonist tinge'.[52] For Plato, the image of eternity is just time as a whole, graspable in thought but given only as a becoming.[53] Doing whatever gives one the 'highest feeling' is apparently what will impress eternity on what would otherwise be only a fleeting life. This is puzzling, but its identification with living 'in such a way that you must wish to live again' at least makes a link with the eternal perspective.

We still need to find out what gives us this highest feeling. That discovery will be just the experience itself: only then will we know how we 'must wish to live again'. It is tempting to suppose that we can resolve this issue in advance, by reviewing or monitoring the course of our life and assessing our experiences according to this yardstick. That idea has appealed to some because it provides the theory, which would otherwise be classed as only theoretical (or 'cosmological'), with a corresponding application. It gives us something to *do*. And Nietzsche does write: 'Not to look out for distant, unknown blisses and blessings and mercies, but to live so that we want to live once again and want to live so for eternity! Our task confronts us in every moment.'[54] But talk of a 'task' needs to be linked with another characterization of the doctrine's bearing on human life. Nietzsche calls it 'a selective principle'.[55] As writers such as John Richardson have noted, he is adapting the Darwinian concept of natural selection for his own purposes.[56] Hence, any suggestion of agency is only metaphorical: nothing is performing a selection. Still less is being selected something that one does, even when it takes place on the basis of what one does. It occurs as a natural process, and I think the same is true of what is called a 'task' in passages like this one.

Nietzsche wants to retain the advantages of the religious attitude for life without the downside of its denial of the world of becoming in favour of an imaginary 'real world' of being. Not all the possible consequences of achieving the eternal perspective are harmful. There is another alternative, one that coincides with the 'metamorphosis' theme at the beginning of *Thus Spoke Zarathustra*. This is the remaining possibility for the impact of the message of eternal recurrence. Nietzsche says that it would 'change you as you are'. Change into what? We are not told. Many readers turn to 'On the Vision and Riddle' for an answer. After all, the doctrine is openly stated there, and one might expect this event to have its own consequences. Our reading of the chapter has not yet reached this point. With that in mind, the next section will discuss its ending, the further narration in which the vision and riddle of the title are finally presented to the reader.

The Shepherd and the Snake

The ending of 'On the Vision and Riddle' needs to be placed in its context. Taken as a whole, the chapter is difficult to follow because its published version has been put together from a mixture of texts. The original plan was considerably altered by a lengthy interpolation which, unfortunately, prevents most readers from noticing the link between the beginning and ending. Commentators then assume that the vision and riddle are found in Zarathustra's account of his polemical confrontation with the spirit of gravity, at first on a mountain path and then at a gateway.[57] In the text, however, these events are presented in a straightforward narrative, and indicated as recent occurrences. There is no vision here, and no riddle – at least, no riddle addressed to the audience identified at the beginning of the chapter, as distinct from the riddle-like question that Zarathustra puts to the spirit of gravity in his first challenge.

In this light, it is worth imagining the chapter without the episode involving the spirit of gravity. If we omit that passage, what remains? A shorter and more unified narrative, for a start. To begin with, Zarathustra is making a sea voyage. He is 'cold and silent', giving no response to glances or questions. He spends several days more or less sulking in his cabin. Eventually he emerges to break his silence with a dramatic narrative, the one we have been reading. In the published text, the ending of this episode is signalled by the sound of a dog howling, 'in the stillest midnight when even dogs believe in ghosts'. The ghost here is the moon, pausing as it passes over a flat roof. Now the scene changes again, and Zarathustra describes an encounter with a young man in desperate distress.

> Among wild cliffs I stood suddenly alone, bleak, in the bleakest moonlight. *But there lay a man.* And there – the dog, jumping, bristling, whining – now he saw me coming; then he howled again, he *cried.* Had I ever heard a dog cry like this for help? And verily, what I saw – I had never seen the like. A young shepherd I saw, writhing, gagging, in spasms, his face distorted, and a heavy black snake hung out of his mouth. Had I ever seen so much nausea and pale dread on one face? He seemed to have been asleep when the snake crawled into his throat, and there bit itself fast. My hand tore at the snake and tore in vain; it did not tear the snake out of his throat. Then it cried out of me – 'Bite! Bite its head off! Bite!' Thus it cried out of me – my dread, my hatred, my nausea, my pity, all that is good and wicked in me cried out of me with a single cry.

Zarathustra pauses and invites his audience to interpret his vision and, in particular, to identify the shepherd into whose throat the snake has crawled. Without waiting for a response, he continues with his narrative. The shepherd does bite the head off the snake and spits it out. Then he jumps up, transformed into a more-than-human being, joyous and laughing. His laughter has a strange effect on the narrator. It fills him with a desperate longing. Zarathustra concludes in renewed agitation: 'My longing for this laughter gnaws at me: oh, how do I bear to go on living! And how could I bear to die now!' So the chapter ends.

Much of this should be familiar to anyone who knows *Tristan und Isolde*, always far more important for Nietzsche than Wagner's other music-dramas.[58] At the start of the chapter, Zarathustra performs a close parallel to the opera's first act. As the curtain rises, Isolde is discovered travelling by sea from Ireland to Cornwall. During the first days of the voyage she has been silent and cold, preoccupied with her own thoughts, but this mood gives way first to a sudden, desperate outburst and then to a lengthy monologue. In her narration, Isolde describes how she came to the aid of a young man in desperate distress, and how her act of compassion had catastrophic consequences for her own well-being. The theme of unsought insight and sudden decision is repeated within her narration, as it is in Zarathustra's. Both describe themselves at one point as moved by an irresistible impulse to cry out: 'A cry came from deep within me!' says Isolde, and Zarathustra repeats this almost word for word. In each case a reversal of fortune, and more particularly of a relation of power, is the story's crucial dramatic feature. Isolde finds herself under Tristan's authority and treated with insulting condescension by him and his supporters. Zarathustra sees the young shepherd transformed into a radiant figure, something higher than human, and ignoring his benefactor – and is made sick with longing by the sight. The attitude of the transformed shepherd is indifference rather than contempt, but its effect is not so different. His laughter celebrates his escape from a world in which Zarathustra is still left. 'With such riddles and bitternesses in his heart Zarathustra crossed the sea', Nietzsche concludes.[59] All this accounts for Zarathustra's deep gloom at the beginning of the chapter, just as Isolde's narrative accounts for her bitter mood. In both cases we have come full circle, achieving a greater understanding of the given situation in the process.

Nietzsche uses this story-within-a-story as a framework into which quite different material is added. The lengthy episode involving the spirit of gravity has distracted and confused his interpreters. Many assume that the riddle posed to the sailors is to be found in the section following immediately on Zarathustra's challenge.[60] In fact, it is expressed in his closing questions: '*What* did I see then in a parable? And *who* is it who must yet come some day? *Who* is the shepherd into whose throat the snake crawled thus?' No such riddle is asked by Isolde, and yet even here we find a parallel, for an issue of identity is central to her story. Although the wounded Tristan has given a false name 'Tantris' (sufficient for disguise in an operatic context) Isolde guesses who he is: the enemy of her people and, what is more, her own personal enemy, since he not only killed her intended husband in the fight that caused his own wound, but returned his head to Ireland as a defiant gesture. Moved by pity (and then love, as we soon suspect) she not only heals him but conceals his real name from her people, and so saves his life twice over.

The story of the snake has other origins. It is similar to a parable of the Persian mystic and poet Jalal al-Din Rumi.[61] In Rumi's version, a prince (*amir*) on horseback encounters a sleeping man into whose mouth a snake is entering. Arriving too late to prevent this, he starts striking the sleeping man with his staff, and then forces him to eat rotten apples from a nearby tree until, tired and nauseated, the man vomits out both the apples and the snake. Clearly there are differences in this version: there is no question of any reversal of power, or of ingratitude on the part of the man saved from mortal peril. In any case, whether Nietzsche knew of Rumi's story is unknown. A related snake theme is present in *Daybreak*, occurring in the discussion of eternal punishment already cited. Nietzsche gives three illustrations of the doctrine's harmful effects on believers, one of which is a Christian of the Middle Ages. 'Dreadful portents appear to him: perhaps a stork holding a snake in its beak but *hesitating* to swallow it.'[62] Can we see a relation between this theme and the story of the young shepherd? Many interpreters have taken the snake to be a symbol of the thought of eternal recurrence, primarily because of its 'heavy' and dangerous character, and there are certainly parallels between this theme and the doctrine of eternal punishment. In a later chapter, Zarathustra undergoes a collapse following a confrontation with his 'abysmal thought' that, like the shepherd's state of distress, replicates the effects of what Lecky calls 'religious terrorism' in eighteenth-century England.[63]

'On the Vision and Riddle' concludes on an inconclusive note. Zarathustra now seems to be speaking only to himself, absorbed in the mood of longing that has a deeper resonance with *Tristan*. He echoes Isolde's words in Act Two: 'oh, how, do I bear to go on living! And how could I bear to die now!' He has asked questions that are left unanswered. Commentators search through later chapters to find resolutions, and often claim to have succeeded in tying up the loose ends, a pretension that matches their triumphalist reading of Zarathustra's development in the work. In the ending of 'On the Vision and Riddle', they see the exultant, carefree shepherd as either Zarathustra or their own possible selves, or even both at once. They give little attention to the equally important but more problematical condition of the story's actual protagonist, even though this is indicated as the real issue in the chapter's closing words.

When Zarathustra eventually manages to summon up his abysmal thought in 'The Convalescent', the consequences are even worse, casting doubt on the naïvely expressed assurance of his animals that his mission has been accomplished. Zarathustra's successes are failures for himself. Yet one can hardly deny that, in some sense, there is a strong forward movement to the work.

The deeper themes emerge in the later parts, and the weaker stretches of Part One, where the author indulges his preoccupations (we do not really want to hear Zarathustra's opinions of his friends' marriages) are not paralleled there. On the other hand, Nietzsche's need to use literary styles that are far less straightforward makes the task of interpretation more difficult, quite apart from the elusiveness of the ideas themselves.

When one bears all this in mind, the chapter's conclusion is not without some ambiguities. Most commentators take it to be an inspirational vignette, on the uncritical assumption that the transformed shepherd is intended as a model for readers, inspiring us to redouble our efforts towards self-overcoming. It should not be taken for granted that Zarathustra is eager for his listeners to hit on the right answer. Does he really want them to achieve the same ascendancy as the transfigured shepherd? Zarathustra is a conflicted figure throughout the work, despite his periodic displays of bravado – and, in fact, this complexity makes him more interesting than he appears in most interpretations.

Chapter 8

The Way of Greatness

This book began with Zarathustra's saying: 'He who would learn to fly one day must first learn to stand and walk and run and climb and dance: one cannot fly into flying.' Other chapters have explored the 'ways' – gateway, lanes, and mountain path – corresponding to some of these forms of motion, but we have not yet come upon any way corresponding to dancing or flying. There is such way in *Thus Spoke Zarathustra*, however. It is not found in a single passage but runs through the work, especially its later parts. Its clearest expression is found in the opening chapter in Part Three, 'The Wanderer'. There it is called the 'way of greatness', and that is the label that I will use in the following discussion.

As this chapter begins, Zarathustra is again climbing, and tells himself that he faces his 'hardest way' and 'loneliest wandering'.

> But whoever is of my kind cannot escape such an hour – the hour which says to him: 'Only now are you going your way of greatness! Peak and abyss – they are now joined together. You are going your way of greatness: now that which has hitherto been your ultimate danger has become your ultimate refuge. You are going your way of greatness: now this must give you the greatest courage that there is no longer any way behind you. You are going your way of greatness: here nobody shall sneak after you. Your own foot has effaced the way behind you, and over it stands written: Impossibility.'[1]

Nietzsche scholars have been unable to make much of this passage. Possibly a certain impatience to get to the following chapter, 'On the Vision and Riddle', is at work. They tend to identify the 'way of greatness' with the mountain path that Zarathustra is climbing as he says these words 'to his heart'. Sometimes this path is described as leading to the highest peak.[2] A mountain path is usually a way to travel from one side of a mountain to another *without* climbing the peak, even if it does involve climbing and has a highest point of its own. In contrast, the way of greatness (not 'to' greatness, as in some translations) has no destination set in advance, and is unlike a mountain path in other ways – for example, in leaving no visible trace and in excluding any companion. It is announced by an unnamed 'hour', perhaps the 'stillest hour' of the end of Part Two, although we are also told that the time is 'about midnight'. As will be seen, this connection with the hour is very significant.

The way of greatness is like the course of a ship through the sea or a bird through the air: neither follows a marked route or leaves a trail for others. That is why nobody else can follow the traveller. Whether the sign 'Impossibility' refers to this, or to the impossibility of retracing one's own steps, is left open. At any rate, this way is for one person alone. Emerson makes the same point: 'When good is near you, when you have life in yourself, it is not by any known or accustomed way; you shall not discern the footprints of any other.'[3] Nietzsche often echoes this thought. He writes, 'There exists in the world a single way along which no one can go but you: where does it lead? Do not ask, go along it.'[4] As Zarathustra puts it, '"This is my way – where is yours?" – thus I answered those who asked me "the way". For *the* way – that does not exist.'[5] In contrast, the two lanes that meet at the gateway are permanent thoroughfares, established for public use.[6] Those ways are common to all: 'Must not whatever *can* run have run on this lane before?' Although only a rough track, the mountain path too remains there for other climbers on other occasions. The way of greatness is not shared even to this extent. The lanes are for everyone, the mountain path for some, the way of greatness for one only.

The sequence that proceeds from gateway and lanes to mountain path and to way of greatness is a progress involving a growth in power and autonomy, displayed in the traveller's increasing freedom of movement – from standing to walking, then to climbing and to flying – and implied in its corollary, the decreasing number of travellers on each way, showing a successive selection at work. Those who travel on the lanes of past and future are described as walking or running, and yet they cannot be called originators of this process. They move because they are compelled to do so, being 'knotted together' so that each draws after all it those that are to come. Zarathustra has already said that the will is a prisoner of the past, and this too is an image of captivity. In contrast, one finds a degree of freedom in the model represented by the mountain path, where the traveller overcomes obstacles, resists external forces or inner drives and achieves a chosen end. Although climbing involves effort and struggle, it also promises a measure of control over the course of events. Complete autonomy, however, is attained only with the power symbolized by flying. The overall tendency is thus an increase in concrete individuality, moving towards the self-sufficiency of a complex totality.

The travellers on these ways vary accordingly. The higher men who climb the difficult path are still limited by their own overcoming. They have attained the second of the three stages of development described in 'On the Three Metamorphoses'. Their way is that of the lion, who defies the authority of the moral law, and replaces 'Thou shalt' with its 'I will'. But a freedom from old values is not sufficient for the creation of new values. The will needs to 'unlearn' its heroic striving to attain this higher state of being. Zarathustra's symbol for a new beginning is the child, a carefree and innocent being, unburdened by commitments or responsibilities. The child's activity is a playful gathering and scattering of counters, a game with no governing rule or final goal. The question is whether we can identify this form of life with any 'way'; for it seems that a way without a direction or destination would hardly be a way at all. This is where the 'way of greatness' needs to be explored in greater detail.

The Dynamics of the Way of Greatness

In the way of greatness, we are told, the conflict between peak and abyss is overcome: 'they are now joined together (*in Eins beschlossen*)'. What this implies is a removal of the opposing forces that hinder our movement, confining it to a narrow path under threat of destruction. On the mountain path, conflict was symbolized as between eye and foot, or between glance and hand. Elsewhere Zarathustra warns against not looking where one is going. 'And many of them walk forward while looking backward with their necks stiff: I like running into them. Foot and eye should not lie nor give the lie to each other.'[7] The tension that these images represent is between *knowing* and *doing*. With the gateway and lanes, the two are kept apart: the perspectival standpoint is about knowing, whereas the ways on which motion occurs are about doing (and for that reason, those on the lanes are not to be taken as having a perspective either on the gateway or on anything else). On the mountain path, knowing and doing are brought together, but enter into an irresolvable conflict. To overcome it would require a move to a new setting in which knowing and doing are reconciled. This is just the 'way of greatness'.

The way of greatness is not a straight line, but an irregular and unpredictable course. 'All good things approach their goal crookedly', Zarathustra remarks.[8] Spontaneity, improvisation, an accepting and taking advantage of chance: these are characteristics of the activity that Nietzsche suggests in evoking images of dancing, sailing and flying. The absence of oppositions and tensions, such as that between peak and abyss, is definitive for this picture. According to Zarathustra, 'bird-wisdom speaks thus: Behold, there is no above, no below!'[9] The wisdom of birds is not a theoretical but a practical wisdom, displayed in their ability to move freely throughout space – that is, to fly.

Flying is Nietzsche's most frequent expression of the ideal to which our human forms of movement – walking, running or climbing – aspire. Zarathustra addresses the sky:

> And when I wandered alone, for *whom* did my soul hunger at night, on false paths? And when I climbed mountains, *whom* did I always seek on the mountains, if not you? And all my wandering and mountain climbing were sheer necessity and a help in my helplessness: what I want with all my will is to *fly*, to fly up into *you*.[10]

As earthbound human beings, however, we can attain this state only in imagination or in dreaming:

> Suppose someone has flown often in his dreams and finally, as soon as he dreams, he is conscious of his power and art of flight as if it were his privilege, also his characteristic and enviable happiness. He believes himself capable of realising every kind of arc and angle simply with the lightest impulse; he knows the feeling of a certain divine frivolity, an 'upward' without tension and constraint, a 'downward' without condescension and humiliation – without *gravity*![11]

Flying is the culmination of the advance whose earlier phases have already been identified: standing, walking, running, climbing and dancing. Nietzsche had given thought to this progression for many years. An early version appears in *Human, All-Too-Human,* where walking and dancing are succeeded by a higher state which is there called 'floating' (*Schweben*), but is more often described as flying:

> Someone who has become tired of play, and who has no reason to work on account of new needs, is sometimes overtaken by the longing for a third state which is related to play as floating is to dancing, as dancing is to walking – for a state of blessed, serene agitation: it is the artists' and philosophers' vision of happiness.[12]

The tendency towards greater power and freedom in this sequence is seen in the increasing number of dimensions in which each form of movement occurs. With standing, of course, there is no change of place. For those who walk or run on the two lanes, there is motion in only one direction: these are one-way streets. But a mountain path has two directions. 'The path up and down is one and the same.'[13] At different times, Zarathustra goes in each of these directions: 'he climbed on and on, up and down'.[14] Stairs and ladders, the other correlates of climbing, have the same twofold directionality. Thus, Nietzsche can write of Zarathustra: 'The ladder on which he ascends and descends is tremendous.'[15] Flying, however, is a free motion, in that every dimension of space is a possible course. Hence, flying can go even where there is no identifiable route. 'Neither by land nor by sea will you find the way to the Hyperboreans.'[16] Nevertheless, the hero Perseus travelled to and from the land of the Hyperboreans – by flying, with the aid of winged sandals.

A motion indifferent to orientations other than those defined by itself is capable of infinitely many directions. There is a close link here with infinite space. When Democritus located the world in infinite space, Nietzsche observes, he retained the notion of 'up' and 'down', not realizing how anomalous it was within such a conception.[17] For Nietzsche, the significance of infinitude lies not in unlimited magnitude, but in an absence of all orientation. Infinite space, whose centre is everywhere and whose circumference is nowhere, is as indifferent to position as it is to direction. Since there are no natural places in this space, our own location, as well as our movement, must appear arbitrary. We find ourselves in the predicament communicated by the 'madman' in *The Gay Science*:

> What did we do when we unchained this earth from its sun? Whither is it moving now? Whither are we moving now? Away from all suns? Are we not plunging continually? Backward, sideward, forward, in all directions? Is there any up or down left? Are we not straying as if through an infinite nothing?'[18]

Seen in this way, our situation is a frightening one; but there is another version of the same theme. Elsewhere in *The Gay Science*, Nietzsche speaks of a 'new infinite' on which the thinkers of the future are prepared to embark, to explore the endless possibilities for interpreting the world. 'We have left the land and have embarked. We have burned our bridges behind us – indeed, we have gone farther and destroyed the land behind us.'[19] For the daring 'searchers and researchers' he praises, even the news that 'the old god is dead' comes as an affirmative, liberating thought, marking a new dawn: 'At long last the horizon appears free to us again.'[20]

The ability to shift at will from one perspective to another is symbolized by movement which is free and unlimited. The search for knowledge requires the greatest possible mobility for the knower. Anyone confined to a single place, like the prisoners in Plato's cave, will be unable to gain adequate knowledge. So too, someone who can move only in a single direction, like the travellers on the 'lanes' of past and future, will be unable to grasp time other than as having a single direction. Yet Nietzsche suggests that 'some beings might be able to experience time backward, or alternately forward and backward (which would involve another direction of life and another concept of cause and effect)'.[21] As we saw in Chapter 2, anomalies of time are already present in our own experience. The need to have a sufficient reason for every experience (for instance, to assign a motive to every action) gives rise to a construction of imaginary causes, projected as events occurring earlier in time.[22] As things are, these habits of thought reinforce our constraint to a single perspective. Yet recognizing them gives us a glimpse of other possibilities of experience. A complete freedom of movement would open up infinitely many perspectives; and this is what 'the power and art of flight' offers for the knower.

There is one human activity that provides our closest approach to this ideal: dancing. Zarathustra says: 'A man's stride betrays whether he has yet found *his* way: behold me walking! But whoever comes close to his goal dances.'[23] He tells his companions that they have not yet reached this stage: 'You higher men, the worst about you is that all of you have not learned to dance as one must dance – dancing away over yourselves!' Dancing with light feet, spontaneous and graceful, presages a liberation from the constraints of gravity. It must, however, be seen in its relation to other forms of movement. Standing, walking and running are commonplace, everyday activities. Climbing, however, involves danger, and therefore requires courage, which 'slays dizziness at the edge of abysses'.[24] It is heroic and, to that extent, *übermenschlich* rather than human. The spirit of gravity lacks this courage, and so cannot even stand on the height; nor can he answer the riddle posed by Zarathustra. Dancing belongs to a higher order of rank again: it is a playful activity, not concerned with overcoming obstacles or dangers, and not aiming at some goal outside itself. The tension and effort involved in the drive to overcome difficulties are no longer present. Dancers have 'light feet'. Not courage (*Mut*) but exuberance (*Übermut*) is what characterizes their activity.

Dancing symbolizes a condition in which certain accepted divisions and oppositions are eliminated. The first of these is the separation of soul and body, a main target of criticism in the First Part of *Thus Spoke Zarathustra*. This distinction, Zarathustra asserts, is a rationalization for the rejection of the body.[25] In identifying ourselves with an immaterial soul, we devalue both the body and the world to which it belongs. Yet the true self, he asserts, is just the 'creative body', which is far wiser, more powerful and creative than the spirit it has created as its instrument. Dancing affirms the body, or rather, in dancing, the body affirms itself and its world, rather than seeking compensation in some 'afterworld'. There is a role for thinking in this, but it is a subordinate one. Even more important for the significance of dancing is the distinction between subject and action. One aspect of dancing, for Nietzsche, is its realization of just this condition. Dancing as he understands it is spontaneous and improvisatory, the expression of a creative impulse, in a process of sustained intensity and continuity. For that reason, it provides no pretext for positing a subject raised above the course of becoming or having an independent status.

The way of greatness is described in mainly negative terms. It is travelled by only one person, is not marked out in advance and, like a bird's flight through the air or a ship's course through the sea, it leaves no trail for others to follow. But it also has a positive character, defined by spontaneity and self-expression. Its forms of movement, dancing and flying, are symbols for the achievement of freedom and power. The precondition for these qualities is a resolution of the conflicts of force which were characteristic of the previous ways. Those tensions acted as restrictions on movement, which was confined to one direction on the lanes, or two directions on the mountain path, or prevented from occurring at all in the gateway. Only an equilibrium of dominating forces creates the possibility of freedom for the more subtle influences that determine what happens on the way of greatness.

A Different Mode of Time: The Hour

Can we identify a temporal meaning for this way, as for Zarathustra's other ways? A direct hint is found in 'The Wanderer': Zarathustra speaks of the *hour* which tells him: 'Only now are you going your way of greatness!' The 'hour' figures prominently in various parts of *Thus Spoke Zarathustra*, and has a special meaning there. It is important to distinguish this use from everyday language, where (since the invention of mechanical clocks, at any rate) an hour is just a certain period of time, like a day or year. These are all parts of time, as Plato puts it, but the 'hour' that figures in *Thus Spoke Zarathustra* is something different. It is one of Nietzsche's most important insights, and the culminating point of his thinking on time and becoming.

Nietzsche makes frequent use of the idea in *Thus Spoke Zarathustra*. Some hours appear in person, while others are referred to or prophesied as still to come. The hour usually has a distinctly individual character, exemplified in the 'hour of great contempt', the 'stillest hour' and the 'blessed hour'.[26] Later, in *On the Genealogy of Morals*, we find the 'golden hour' and the 'reflective hour'.[27] Other hours are described but not named. Noon and midnight are important hours, often referred to in the work. The 'hour of noon'[28] suggests Zarathustra's prophesied 'great noon'.[29] Hours are personal, not common to all: Zarathustra speaks of 'my hour', just as he speaks of 'my way'.[30] Thus, the hour shares many features with the way of greatness. But to determine what form of time is characterized in the hour, we need to pursue this comparison into aspects which have some temporal meaning.

Whereas the two lanes stretch into infinite distances, the 'way of greatness' is confined to a single location, the spot where the traveller's foot is set down. This is not an unextended point, but a place occupied by a living being. Similarly, the 'hour' is a living present, not an instant. Hours do not pass by in the glance of an eye: they are times within which reflections, conversations, and other events can take place. They arrive and depart, and so have beginnings and endings, like ordinary intervals. It is what comes between these boundaries that is puzzling.

Hours seem to have a certain timelessness. That is, they are not to be characterized in terms of ordinary duration, as measured by clocks: 'All this lasted a long time, or a short time, for properly speaking, there is *no* time on earth for such things.'[31] The time of the hour is not commensurable with the time of the world outside it. We find echoes of this in a prose poem of Baudelaire, describing a vision of time within the eyes of his cat Féline: 'always at the back of her adorable eyes I can distinctly see the time, always the same – vast, solemn, wide as space, without minutes and without seconds – a motionless hour not marked on any clock, and yet as airy as a breath, as quick as a glance'.[32] This is not timelessness in the sense of an absence or negation of time. If the time of the hour is elusive, that is only from the standpoint of 'time on earth'. What occurs within the hour has its own time, even if it belongs to a human life. In that respect, the hour is always 'untimely'. It occurs when it is 'high time'[33] or 'the right time'.[34] This is the meaning of the Greek word *kairos* in statements attributed in the Gospels to Jesus, such as 'You do not know when the hour will come' and 'My hour has not yet come' (a statement echoed by Zarathustra, but rescinded in Book Four's conclusion).[35] No clock can tell us when this 'right time' is, since it is not constituted by its relation to other times, as periods of world time are.[36] Rather, it is determined by the specific character of the hour.

Yet speaking of what occurs 'within' the hour is misleading, if it implies the distinction between time and things in time – between the form and content of temporality – that was a feature of the previous ways. For the gateway and lanes, the distinction is between the ways themselves and those who walk or run on them. Similarly, a separation of locations and events or states of affairs is presupposed in the schematism of finite intervals. The term 'recurrence' usually means an occurrence of the same event or process at a different time. This implies that the event and the time at which it occurs are distinct from each other and, indeed, indifferent to each other. On that assumption, there can be no question of a 'right time': one time has the same status as any other. In contrast, Zarathustra describes the 'great noon' as having 'its time and its own destiny (*Schicksal*)'. What is meant by 'destiny' is just the necessary unity of the two, determined not by some external agency (such as so-called Fate) but by their own nature.

Nietzsche describes modern nihilism as a state of homelessness, at least for those who understand its implications. Regarding oneself as homeless, he suggests, is distinctive and honourable for some people, an entitlement for the free spirit: 'We children of the future, how *could* we be at home in this today?'[37] The analogy is clear: a 'right' time would correspond to a 'right' place – that is, a home. But what does a 'right' time have to do with past, present and future? At first sight, Nietzsche seems to be saying that 'we' would be happier living in some future age, in which the quality of life might suit us better. Other people may be attracted by some past period or other; understandably, they too will feel just as homeless in today's world. Whether the century one lives in coincides with one's preferred time is, it seems, a matter of chance.

This is a ready interpretation which, I think, fails to catch the subtlety of Nietzsche's thinking. Once again, the relation between past and future is confused with that between earlier and later. In these instances, one present time is being compared with some earlier or later present time. Now, this may seem the only relation to be considered here. After all, if whatever occurs does so in the present, the question of living in the past or future does not arise, except in the sense of later or earlier 'nows'. But Nietzsche is speaking of past and future in a way that is less straightforward. He describes 'untimely' persons as belonging to 'tomorrow and the day after tomorrow'.[38] Inevitably, they feel out of place in the world they live in. This does not mean that their problem could be solved by using a time machine to travel to a later time. Despite what seems a natural mode of expression, nobody can move into the future: the time traveller will still find himself living in the present, whatever his eventual location in time.

Zarathustra's challenge to the spirit of gravity asserts that the present moment must occur 'out there' where 'all things' run into infinite distances, and suggests something like an infinite proliferation of the moment through past and future expanses of time. The 'way of greatness' proposes a very different process. Near the beginning of 'The Wanderer', Zarathustra says: 'What returns, what finally comes home to me, is my own self and what of myself has long been in strange lands and scattered among all things and accidents.' Here what occurs within the hour is given a more concrete description, as a return or, more concretely still, as a coming home. Werner Stegmaier argues that a 'right time' is always still to come and never present.[39] This may fit his example, Zarathustra's 'great noon' but, as other instances show, what I am calling the 'hour' figures in *Thus Spoke Zarathustra* not only as a prediction but as an actuality.

For a further clue, we should return to Emerson, who praises the simple mind which 'lives now, and absorbs past and future into the present hour'.[40] Near the end of Part Three, Zarathustra says to his own soul: 'O my soul, now there is not a soul anywhere that would be more loving and comprehending and comprehensive (*umfangender und umfänglicher*). Where would future and past be closer together than in you?'[41] This closeness of future and past expresses a feature of the 'way of greatness' already remarked on: the absence of tensions and opposing forces. It is not that these forces have been eliminated – that would be impossible – but rather that they have been incorporated by the traveller. 'Incorporated' means, literally, taken into one's body; but remembering that, for Nietzsche, the body *is* a field of forces, this is just what he has in mind. Incorporation brings about an equilibrium of force. As Emerson observes in his essay 'Fate', we are not crushed by the weight of the atmosphere, because the same force is present within our bodies and compensates for the pressure from outside.[42] Instead of excluding past and future, the hour draws them into itself, so that they no longer oppose each other, as in the gateway 'Moment', but rather dwell together. If a present is caught between these opponents and is unable to withstand them, it is reduced to the instantaneity of the moment. The hour prevents such an outcome by encompassing past and future, by providing a place for them to dwell together; and so it attains a balance and stability which is missing in the transitory moment.

Is this the same as Arendt's concept of 'thinking', discussed in Chapter 5? That is presented in *The Life of the Mind* within a Nietzschean context, as a liberation of the individual person from oppression by the forces of past and future. There are seeming parallels, especially when Arendt asserts that thinking 'gathers the absent tenses, the not-yet and the no-more, into its own presence'.[43] But her emphasis is on the quiet stillness of meditative thought, and use of expressions such as 'non-time' and *nunc stans* (also invoked by some Nietzsche interpreters, it must be allowed) signal that becoming itself is supposed to be suspended in this 'gap' between past and future.[44] In contrast, the hour is timeless only in relation to the time of the world. It has a time of its own, and things happen there, including dramatic encounters. Accordingly, past and future are not just represented in the hour, but figure there as themselves – without that, becoming would not appear as time.

This absorption of past and future within the present hour is a homecoming for the dispersed self. A homecoming, however, is not an arrival at some new and unfamiliar place. Is it possible that we already have an experience of this form of temporality? In a draft version of his 'untimely meditation' on history, Nietzsche describes our feelings of envy in looking on the animals who live wholly within the present, or on the child who 'plays in too short and blissful blindness between the two fences (*Zäunen*) of past and future'.[45] Later the child will learn the meaning of regret, guilt, hope and fear – all not very successful attempts to come to terms with the claims of past or future over the present. When that occurs, its play area is overrun by the forces that had been kept outside as long as the expression 'It was' had no meaning. With the elimination of this space there is left just a single gateway, within which past and future come together and meet face to face, as we have seen. While the child's state of innocence lasts, however, it is a naïve version of the later 'way of greatness'. Hence, the parallel between this original experience and its regained counterpart is quite close.

In the hour, then, past and future exert no power over the present. That is not to say that the past and future outside the hour are abolished: if they were, the hour would stand outside time, and those concepts would have no significance for it. The analogy between the time within the hour and the space within an individual living body is again relevant. In both cases, there is a contrast between what might be called 'internal' and 'external' points of view. For other people, the space occupied by my body is like any other space. But insofar as every part of my own body is immediately present for me, there is, in a certain way, *no* space inside my body. As for the space that begins at its boundary, this is a space of different directions by virtue of its relation to my bodily form. It is before and behind me, left and right, and so on; in fact, it has infinitely many directions. The declaration of Zarathustra's 'bird-wisdom', 'Behold, there is no above, no below!' seems to deny any directionality at all; but what it rejects are the directions that are defined by reference to the external world, and determined by forces acting upon the traveller. Above and below, in this sense, are crucial directions for another 'way', the mountain path. There the tension between height and depth requires the traveller to keep to a narrow track, along which movement can only be an ascent or descent. An imposition of directions, however, is just what bird-wisdom repudiates when it asserts a right to determine its own orientation. The equilibrium of external forces opens up an opportunity for more subtle forces, suppressed by the familiar 'ways', to express themselves in free and spontaneous activity which, in turn, gives rise to infinitely complex and unpredictable forms.

These features of the hour explain why it is Nietzsche's ultimate temporal paradigm. The first departure from pure becoming towards determinate models of time was a separation of the form and content of temporality. In the hour, these alienated elements are reunited, but in a way that can be grasped. The hour is thus a higher realization of becoming, incorporating the previous stages of development while being something radically different. Of all Nietzsche's temporal concepts, this is the most neglected by scholars. Worse still, a great deal of confusion is caused by running it together with the 'moment'. It may be said that in the hour, time no longer exists, or that time stands still, as the German word *Stunde* suggests. But this is true only of the common time of the world. Within the hour there is becoming and scope for things to happen: meetings, conversations and other events. Thus, the hour cannot be identified with a point of time, even if both are contrasted with the periods occupied by events and processes.

The hour has a beginning and ending, and therefore a relation to the time that lies outside it, although the two are disparate and mutually incommensurable. One passes from mundane time to the hour – for example, the 'stillest hour' begins 'towards evening' – and back again once the hour has reached its completion. This suggests a more general thesis: that within the scheme set out in previous chapters, there are transitions between any stage or level and the next. Zarathustra is climbing the mountain path when he comes to the gateway and lanes; and after the encounter at the gateway he finds himself 'among wild cliffs', suggesting a return to the mountain scene. Again, the travellers who enter the gateway from one lane do so only to leave it on the other lane. So it appears that, despite the differences between the ways, there are passages between them and between the corresponding modes of temporality.

'This Life – Your Eternal Life!'

Having identified the 'hour' and placed it within the context of Zarathustra's four ways, I want to pursue the theme further by relating it to the thought of eternal return. This is quite difficult, and I will begin by drawing attention to one pitfall. What is it that is supposed to return eternally? As we saw in Chapter 6, Nietzsche gives no single answer to this question, and usually talks in general terms of states of affairs, events, situations, and so on. Yet what he says often seems to support the primary status of the *moment* in relation to eternal recurrence. The 'demon' who announces the thought of eternal recurrence in *The Gay Science* emphasizes the present moment ('and even this moment') and the word's Goethean resonance is reinforced by the question about 'a tremendous moment' and Nietzsche's occasional use of expressions like 'highest moment' elsewhere.[46] Zarathustra confronts the spirit of gravity in the gateway 'Moment' and poses a question about the present moment's recurrence.[47] Given all these cues, it is not surprising that the moment figures as the central element in many interpretations of the doctrine.[48]

Even so, I think it is wrong to understand eternal recurrence as an apotheosis of the moment. The moment symbolized by the gateway is defined by conflict, exclusion and distance, and cannot perform the task of reconciling and integrating the estranged forms of time. In the critical literature, the word 'moment' tends to mark an outbreak of vagueness and ambiguity. On some occasions, though, what is meant may be what I am calling the 'hour'. After all, both are contrasted with shorter or longer periods of time, and to that extent might be regarded as standing outside time. Hence, the same term might be used in both ways. Yet despite having incommensurability in common, their relations to time differ sharply. The instant is constituted by an exclusion of past and future, while the hour is supposed to incorporate them.

This raises the question of the hour's relation to the eternal recurrence. Can the hour be understood as something that returns eternally? None of the 'hours' that figure in *Thus Spoke Zarathustra* seems to make more than one appearance, but that does not answer the question, since recurrence within a particular life is not the issue. The hour might be relevant in a different way, if we consider a human life taken as whole, and try to relate that both to this mode of temporality and to the eternal return. There is some justification in the demon's opening statement that what is to recur is 'this life as you now lived it and have lived it', although the part of one's life that is still to come is not included there (unless this encounter occurs at the moment of one's death, as Loeb suggests). A clear indication that what recurs is life as a whole appears in this 1881 notebook entry:

> You think you have a long wait until rebirth – but do not deceive yourselves! Between the last moment of consciousness and the first appearance of the new life lies 'no time' – it goes by as fast as lightning, even if living creatures measure it in billions of years or could not measure it at all. Timelessness and succession are compatible with each other (*vertragen sich miteinander*) when the intellect is absent.[49]

What Nietzsche is asserting looks paradoxical, as his last sentence acknowledges. It is natural to suppose that the time of the world as a whole continues as usual between one's death and rebirth, with a succession of events that living creatures can perceive and measure according to the various standards of their life-processes. Yet elsewhere in the same notebook Nietzsche emphasizes that succession is constructed by the human intellect which posits 'gaps between things'.[50] His claims about the anomalous experiences of time that result from temporary suspension of the intellect's activity are paralleled in his argument here. This succession of events must be a timeless one for the individual who, to express a living observer's standpoint, is 'awaiting' rebirth. The underlying idea can be found in Aristotle's *Physics*:

> But neither does time exist without change; for when the state of our own minds does not change at all, or we have not noticed its changing, we do not realise that time has elapsed, any more than those who are fabled to sleep among the heroes in Sardinia do when they are awakened: for they connect

> the earlier 'now' with the later and make them one, cutting out the interval because of their failure to notice it.[51]

Nietzsche speaks of two particular 'nows': the last moment of the old life and the first moment of a new one. Is he saying that they are the *same* moment? If so, we are back to the notion of closed or circular time. Paul Loeb defends this view and makes a comparison with midnight, taken as the end of one day and the beginning of another: 'Just as the hand of a clock crosses a single unique moment that is at once the last and the first moment, so too Zarathustra experiences a single unique moment in which he is at once dying and awakening.'[52] This analogy contains puzzles, however. Is there in fact a moment which is the last moment of one day and also the first of another? Such a twofold identification is quite problematical. Again, does Zarathustra actually die, or only have a near-death experience? If he does die, then when is he dead? Apparently at the moment of midnight, although not before or after. In that case, he can be having no experiences at that moment, either of dying or of awakening. Moments of stopping and starting present conceptual issues at the best of times, and even more so if the two kinds of transition are supposed to be combined. And certainly the experiences of the dying person will be very different in kind from the first awareness of a newborn infant.

These puzzles make a 'single unique moment' of dying and awakening look less plausible. The reason is not that closed time is incoherent – a claim rejected in Chapter 6 – but that the radical character of death and rebirth is missing from this picture. In a circle there is no beginning and ending, and so a lifetime that forms a closed figure contains no real birth or death. Yet Zarathustra's animals are clear on the specificity of such events: '"Now I die and vanish," you would say, "and all at once I am nothing."'[53] They add that this passing away includes soul as well as body, as Zarathustra had earlier said to the dying tightrope walker. Thus, the immediacy of the individual person's return to life cannot be treated as a straightforward continuity. We need to make another effort to understand this puzzling text.

Nietzsche says that for the human intellect, succession is incompatible with timelessness, and that is why this combination is hard to grasp. What succession does he have in mind? Perhaps of two moments, the last of one life and the first of another, such that one occurs after the other but with no time between them. That would be inconsistent with the Boscovichian principle of continuity, which says that motion from one place to another must pass through all the places in between. Instead, the moments would be like the points in Nietzsche's time-atom theory, which are not separated by a continuous time.

I suggest another identification of the succession occurring in 'no time': that it consists in the many events occurring in the world between the ending of one life and the start of a new life. This is what 'goes by as fast as lightning', although it takes a long time by the measure of living beings. The word 'it' is tied to the phrase 'no time' ('*keine Zeit*'). What is this 'no time' (or 'not a time') and why is it between inverted commas? These may signal some special sense, or an allusion to another writer, or both.[54] Clues to its meaning can be found in Nietzsche's other uses of the phrase.[55] Most relevant is an entry in the same notebook about the part of night 'where there is no time'.[56] As we saw in Chapter 2, that passage refers to a suspension of time which is not timelessness in the sense of a negation or denial of time, but a glimpse of the 'time-chaos' out of which time originally emerges. When the form- and rhythm-shaping function of the intellect is out of action, what remains is a fluctuating becoming of intensities, not yet organized as a succession. Nietzsche is anticipating in calling this 'time-chaos', a concession to the limited resources of language: the content (which is also the form) of absolute becoming is a 'more and less', not a 'before and after'. The image of the lightning flash is the same one he uses when drawing attention to anomalies of time order within ordinary experience: he typically says that they involve 'a hundred details that pass like lightning'.[57]

In this light, the 'no time' of the transition between one life and another must be seen as quite different from a simple negation that results in a picture of circular time. What alternative model is available? We began with the idea that what recurs eternally is a human life. When Nietzsche writes, 'This life – your eternal life!'[58] he is not observing a fact but expressing a vision, at the 'tremendous moment' in which the eternal return is affirmed and the individual's life is grasped as a recurring lifetime, apart from which there is 'no time' in the sense I have explained. This life begins again after death without any intervening interval, even if the course of the world continues in the meantime. The implication is that the time of that life is not just one part of world time, but a time in its own right.

Now, another 'no time' appeared earlier in this chapter in connection with the 'hour'. It is impossible to say whether the hour lasts a long or a short time, Nietzsche says, since 'properly speaking, there is *no* time on earth for such things'.[59] This point may seem separate from the assertion that there is 'no time' between death and rebirth. In fact, the two go together. Just as the hour is 'no time' by the standard of mundane temporality, so the events outside the hour are 'no time' in relation to the hour's own time. This last point, however, recalls the 'no time' that separates one lifetime from another. The question now is, can the hour provide the model we are looking for here?

One of the few thinkers who have presented the hour in some detail is Franz Rosenzweig. In some ways his conception is related to Nietzsche's, although in others it is different. For example, both make use of the suggestion of 'standing' in the German word *Stunde* as an important clue, pointing to a suspension of transitoriness. In Rosenzweig's view, this is closely linked with repetition.

> Such a stationary Now is called, in contrast to the moment, hour. Because it is stationary, the hour can already contain within itself the multiplicity of old and new, the fullness of moments. Its end can merge back into its beginning

> because it has a middle, indeed many middle moments between its beginning and its end. With beginning, middle and end it can become that which the mere sequence of individual and ever new moments never can: a circle returning in upon itself.[60]

What is important for the analogy is Rosenzweig's observation that the hour has a beginning, middle and end, and that it is because of this that we can think of it as recurring. (His paradigm case would be a regular religious observance, such as the Jewish Sabbath.) This is just where a preoccupation with the 'moment', taken by itself, leads to difficulties of interpretation. A telling instance occurs in an insightful discussion of Nietzsche's ideas on time by Ofelia Schutte:

> With regard to his conception of time, Nietzsche represents time not as an undifferentiated succession of moments but as moments differentiated amongst themselves by a movement of parting and coming together. A parting moment may be conceived as being met, or encountered, by an arriving moment in a spatiotemporal rather than merely temporal-sequential configuration of events. Moreover, since whatever parts and arrives in this way is a moment, the 'arriving' moment may also be conceived at a different level of abstraction as a 'returning' moment. If a moment can be thought of as having an end, under 'willing backwards' the end would become its beginning while its beginning as conceived according to linear time would become its end, and so on.[61]

The problems in this line of thought come out in the final sentence. If a moment could have a beginning and an end, Schutte says, then we might imagine moments to be joined together, and the outcome would fit the metaphor of eternal return. That sounds promising, but the hypothetical wording of the conclusion shows that it is really a counterfactual proposition. The assumption that time consists of moments, whether differentiated or not, is unsustainable. It is just the absence of any such distinction within the moment that gives Zeno's 'arrow' paradox its force. The word 'moment' would have to be given a completely different sense to support this line of thought. It would, in fact, have to stand for what I am calling the hour.

The hour provides a model of temporality that is different from the concept of circular time discussed in Chapter 6, because it incorporates the elements of past and future which are crucial for human experience. Further, it has a beginning and ending, unlike a circular figure. Hence, within the hour past and future dwell together rather than running off into endless distances, like the infinite perspectives of the gateway 'Moment'. In some ways it is evident that a human life fits this model well. For example, the personal character of the hour corresponds to the uniqueness of an individual lifetime. The future and past of an individual life are finite, unlike future and past time in general, and that is also true of the hour. This finitude does *not* imply a unique first or last moment, but it does imply a horizon in either direction that rules out, for one thing, the sort of recollection of experiences from past lives that Loeb postulates. In describing the soul's incorporation of future and past Zarathustra uses two closely related words: *umfangend* and *umfänglich*, together implying an inclusiveness which is both extensive and complete. The soul takes in the entire future and past, if not of the world as a whole – an impossible task – then of its own participation in time as a living being.

In earlier chapters we encountered alternatives: the child who simply lives within the present, the metaphysical thinker who rejects temporality in favour of an imaginary eternal reality, and the 'untimely' person who could be said to live in the future rather than the present.[62] The hour is very different from this last, but it can be called 'untimely' in another way, since it is not a part of world time, bound up with whatever lies outside it, but has a completeness of its own. In this and other ways, the concept provides a way of grasping a human life, taken as a whole, that in turn throws light on important aspects of the thought of the eternal return.

The Dance of Gods

Even so, there is one conspicuous difference between the hour and a typical human life. Past and future are usually experienced by us not as dwelling together in harmony but, on the contrary, as in continual conflict. Hence, Zarathustra's description of the gateway 'Moment', intended to be accepted as an accurate picture of human life. As long as this antagonism persists, there is no possibility of seeing one's life as a whole where past and future 'dwell together'. Rather, as Nietzsche puts it in *On the Uses and Disadvantages of History for Life*, life always remains an 'imperfect tense'. Death does not resolve this tension but only 'sets the seal on the knowledge that existence is only an uninterrupted has-been, a thing that lives by negating, consuming and contradicting itself'.[63]

Can we be more specific about what is needed to resolve the contradiction? It is not an abolition of past and future, but a different way of grasping them. As I noted in Chapter 7, Zarathustra not only states the predicament due to 'time and its "It was"' but points towards a solution when he speaks of turning 'It was' into 'But thus I willed it' and 'But thus I will it; thus shall I will it'. In this way, the will is to achieve 'reconciliation with time and something higher than any reconciliation'. Just how such a transformation can be achieved, and how it is linked with the eternal return, continues to be disputed among Nietzsche's interpreters and needs a separate investigation of its own. Here we can only form an idea of what the outcome might be.

I suggest that Nietzsche's concept of the hour is his clearest expression of this possibility. However, it needs to be emphasized that the hour is not a model of time that one can simply decide to adopt. The hours that appear in *Zarathustra* arrive and depart of their own accord. Sometimes they announce themselves, but on other occasions they enter 'on light feet'. In contrast, Rosenzweig's 'hour' is constituted by human beliefs and practices, and marked out accordingly. 'Only the stroke of the bells establishes the hour,' he writes, 'not the ticking of the pendulum.'[64] The striking of the midnight bell in *Zarathustra* suggests something similar, but Nietzsche does not regard the hour as a human creation, or even as grounded in naturally occurring cycles. His bell only marks the arrival of the hour, and does not 'establish it' in Rosenzweig's sense, by bringing a community together for a shared participation. It is not our inventive talent that enables a reconciliation of past and future, but a state of force that expresses itself in a relatively stable form of life, together with a characteristic temporality. If this seems to place the prospect of reconciliation out of reach, so to speak, that is true only for an approach still dominated by the assumption of a controlling subject with freedom of choice, a prejudice that makes Nietzsche's thinking impossible to appreciate.

When the play of forces gives rise to a state of completeness and harmony, there is no implication of permanence: only the impossible state of rest could have that consequence. Nietzsche speaks of 'my Dionysian world of the eternally self-creating and eternally self-destroying'.[65] It is not a world of being but one of continual becoming, in which conflict and concord alternate without end. In contrast, the ancient Greek concept of *kosmos* implied a world bound together, in Plato's words, by 'communion and friendship and orderliness and temperance and justice'.[66] This may be why Nietzsche at times puts inverted commas around the word 'world' when using it in his own sense.[67] Zarathustra expresses this conception when he speaks of being carried away by his longing from the here-and-now to distant places:

> Where all becoming seemed to me the dance of gods and the prankishness of gods, and the world seemed free and frolicsome and as if fleeing back to itself – as an eternal fleeing and seeking each other again of many gods, as the happy controverting of each other, conversing again with each other, and converging again (*Sich-Widersprechen, Sich-Wieder-hören, Sich-Wieder-Zugehören*) of many gods.[68]

The Dionysian conception involves cycles of becoming which pass through distinct and contrasting phases. When hinting at the doctrine in an early passage of *The Gay Science* (well before the appearance of the 'demon' who confronts the unwary reader) Nietzsche speaks of 'the eternal recurrence of war and peace'.[69] Nietzsche's attempts at constructing an *argument* for a cyclical cosmology leave out this central feature. As we saw earlier, that line of thought rests only on the finitude of the states that make up the total process. It is this indifference to content, rather than the logical issues seized upon by interpreters, that is the argument's main limitation.

In an important sequence of notebook entries dating from 1887, Nietzsche adapts a traditional symbol to characterize the Dionysian world of harmony and conflict. He writes, 'The sole possibility of maintaining a meaning for the concept 'God' would be: God not as a driving force, but God as a *maximal state*, as an *epoch* – a point in the evolution of the will to power by means of which further development just as much as previous development up to him would be explained.'[70] Any attribution of moral perfection to God is repudiated as an aberration of the philosophers:

> Let us remove the highest goodness from the concept of God: it is unworthy of a god. Let us also remove the highest wisdom: it is the vanity of philosophers that is to be blamed for this made notion of God as a monster of wisdom: he had to be as like them as possible. No! God the *highest power* – that suffices! From it follows everything, from it follows – 'the world'![71]

This is not a creator who stands over and beyond the natural realm in some eternal 'real world'. Nor is its participation that of a being who is a constant presence, necessary and complete in itself. Rather, it is itself subject to becoming: in other words, it comes to be and passes away. Existence itself, Nietzsche tells us, is 'an eternal god-making and unmaking'.[72] That is, the world may attain a state of necessity and completeness, but it will move onward into a phase of disintegration. The idea of the god, then, stands for this culminating moment, at which the world is grasped as 'perfect'.

The power of Nietzsche's god is not a compelling force, even if a completed world 'follows from' it. In *Twilight of the Idols* he writes, 'All that is *good* is instinct – and hence easy, necessary, free. Laboriousness is an objection; the *god* is typically different from the hero. (In my language: *light* feet are the first attribute of godhood.)'[73] A notebook entry from the period of *Zarathustra* brings out the link between this idea and the Heraclitean image of the child whose activity is characterized by innocence and playfulness. '"Play", the useless, as ideal of whoever is overfull of strength, as "childlike". The "childlikeness" of God, *pais paizon*.'[74] The last phrase cites a fragment of Heraclitus, who in Lucian's only partly satirical survey of contemporary philosophy, 'Philosophies for Sale', is asked the question 'What is the *aion*?' and replies: *pais paizon, pesseuon, sumpheromenos, diapheromenos*: 'A child playing, moving counters, gathering and scattering.'[75] In *Philosophy in the Tragic Age of the Greeks*, Nietzsche had described Heraclitean becoming in similar terms:

> A coming to be and passing away, a building and destroying without any moral valuation, in eternal selfsame innocence, belong in this world only to the play of the artist and the child. And as the artist and the child play, so the eternal living fire plays, builds up and destroys in innocence – and this game the *aion* plays with itself. Transforming itself into water and earth it piles up

sandcastles like a child beside the sea, piles up and tramples down: and from time to time it begins its game anew.[76]

What is described here is not a game that can be won or lost – that is, not a purposeful activity governed by rules and calling for skill, strength or judgement to engage in successfully. Play is one thing and sport is another. For Nietzsche, the hallmark of true play is freedom and creativity: two interdependent aspects of a single form of life. The absence of further ends, along with the authority they impose, is an advantage rather than disadvantage there. The child that gathers and scatters counters with no higher purpose or further aim is 'serious' in this sense, and yet its mood of high spirits is the opposite of the solemnity of the spirit of gravity. The figure of the Nietzschean god shows this childlike innocence and playfulness, but adds an active identification with becoming rather than being. Our question about the hour's departure finds an answer in that participation. This soul does not will to retain its qualities of necessity and completeness. On the contrary, it enters into the process of becoming and abandons itself to chance, disunity and conflict. Nietzsche says that it 'flees itself' – but he adds, 'and catches up with itself in the widest circle'. The high point will be attained again as the circle of becoming is completed. The suggestion here is that any attempt to stop the process, to prevent the highest moment from ceasing to be, would be futile. We can however say to the moment, 'Go, but return.'[77]

Nietzsche's thinking about time reaches its ultimate development in this conception. Having begun with absolute becoming and 'time-chaos', it has covered the ground occupied by our familiar human concepts of time, showing how the conflicts and tensions of becoming give rise to temporal schemata of past, present and future, or of successions taking longer and shorter periods of time. But it has progressively moved towards something higher, to a vision that could be realized fully only if we ourselves were to become something other than we are and have been.

Notes

Preface

[1] Alexander Nehamas, *Nietzsche: Life as Literature* (Cambridge, MA: Harvard University Press, 1985), 150.

Introduction

[1] KSA 8, 7[156], 199.

[2] *Ecce Homo*, 'The Birth of Tragedy', sec. 3.

[3] Aristotle, *Physics* VIII.3, 253b, trans. R. P. Hardie and R. K. Gaye, in *The Complete Works of Aristotle*, ed. J. Barnes (Princeton, NJ: Princeton University Press, 1984), vol. 1, 423.

[4] Volker Gerhardt, 'Die Metaphysik des Werdens: Über ein traditionelles Element in Nietzsches Lehre vom "Willen zur Macht"', in *Nietzsche und die philosophische Tradition*, ed. Josef Simon, Band I (Würzburg: Königshausen + Neumann, 1985), 18.

[5] *Thus Spoke Zarathustra*, 'Upon the Blessed Isles'.

[6] KSA 11, 25[307], 90. Cf. F. A. Lange, *Geschichte des Materialismus und Kritik seiner Bedeutung in der Gegenwart* (Iserlohn: J. Baedecker, 1866), 499.

[7] *Thus Spoke Zarathustra*, 'Upon the Blessed Isles'.

[8] KSA 12, 9[89], 382 (*The Will to Power*, sec. 517).

[9] For a full exploration of this theme, see John Richardson, *Nietzsche's New Darwinism* (New York: Oxford University Press, 2004).

[10] KSA 9, 11[150], 499.

[11] Fr. 12, in G. S. Kirk and J. E. Raven, *The Presocratic Philosophers: A Critical History with a Selection of Texts* (Cambridge: Cambridge University Press, 1957), 196.

[12] KSA 13, 14[79], 258; KSA 13,14[82], 262; and KSA 13, 14[122], 302 (*The Will to Power*, sec. 635, 689 and 625).

[13] KSA 9, 6[433], 309 and elsewhere.

[14] KSA 11, 40[21], 639 (*The Will to Power*, sec. 492). Cf. KSA 12, 2[69], 92 and KSA 12, 7[34], 306; and KSA 13, 14[122], 302 (*The Will to Power*, sec. 625).

[15] KSA 11, 36[23], 561 (*The Will to Power*, sec. 520).

[16] KSA 9, 11[149], 499.

[17] *The Wanderer and His Shadow*, sec. 55.

[18] *Beyond Good and Evil*, sec. 20.

[19] KSA 12, 2[83], 102 (*The Will to Power*, sec. 531).

[20] *Beyond Good and Evil*, sec. 20.

[21] *Human, All-Too-Human*, sec. 19.

[22] KGW II/4, 300, and Nietzsche, *The Pre-Platonic Philosophers*, ed. and trans. G. Whitlock (Urbana and Chicago: University of Illinois Press, 2001), 92. Cf. Nietzsche's letter of August 1866 to Carl von Gersdorff, KSB 2, 160, citing Lange, *Geschichte des Materialismus*, 493.

[23] KSA 12, 2[110], 115.

[24] *Ecce Homo*, 'Why I Am a Destiny', sec. 3.

[25] See e.g. Otto Liebmann, *Kant und die Epigonen. Eine kritische Abhandlung* (Stuttgart: Carl Schober, 1865), 36–9 and 63–9.

[26] *Twilight of the Idols*, 'How the "True World" Finally Became a Fable: The History of an Error'.

[27] *Human, All-Too-Human*, sec. 2.

[28] KSA 7, 26[12], 575–9.

[29] KSA 9, 11[292], 553. See also KSA 11, 36[15], 556 (*The Will to Power*, sec. 1062).

[30] KSA 11, 35[55], 537.

[31] *The Gay Science*, sec. 84.

[32] Pierre Klossowski, *Nietzsche and the Vicious Circle*, trans. Daniel W. Smith (London: The Athlone Press, 1997), 60–1.

[33] KSA 12, 9[91], 384 and KSA 12, 9[62], 369 (*The Will to Power*, sec. 552 and 580).

[34] Plato, *Timaeus*, 37, in *The Dialogues of Plato*, trans. B. Jowett (New York: Random House, 1937), vol. 2, 19.

[35] Cf. Plato, *Parmenides*, 151–2, in *The Dialogues of Plato*, vol. 2, 120, where 'was', 'is' and 'will be' are explicitly tied to past, present and future time.

[36] *Thus Spoke Zarathustra*, 'On the Spirit of Gravity', sec. 2.

[37] *On the Uses and Disadvantages of History for Life*, sec. 1. See the discussion in Chapter 3.

[38] *The Gay Science*, sec. 382.

[39] KSA 11, 38[12], 610–11 (*The Will to Power*, sec. 1067).

[40] KSA 12, 2[110], 115.

Chapter 1

[1] *Philosophy in the Tragic Age of the Greeks*, sec. 10.

[2] Kirk and Raven, *The Presocratic Philosophers*, 117.

[3] KGW II/4, 241, where the fragment appears in Greek. *The Pre-Platonic Philosophers*, 33. He makes the same omission in *Philosophy in the Tragic Age of the Greeks*, sec. 4, following the example of Friedrich Ueberweg, *Grundriss der Geschichte der Philosophie*, ed. Karl Praechter (Graz: Akademische Druck- und Verlagsanstalt, 1953), vol. 1, 48.

[4] KGW II/4, 246. *The Pre-Platonic Philosophers*, 37.

[5] *Philosophy in the Tragic Age of the Greeks*, sec. 4, and KGW II/4, 245. *The Pre-Platonic Philosophers*, 36.

[6] *Philosophy in the Tragic Age of the Greeks*, sec. 5.

[7] Ibid., sec. 5.

[8] Lucian, 'Philosophies for Sale', trans. A. M. Harmon, in Lucian, *Works*, Loeb Classical Library (London: Heinemann, 1919), vol. 2, 476. See KSA 12, 2[130], 129 (*The Will to Power*, sec. 797).

[9] *Schopenhauer as Educator*, sec. 4. Cf. *Philosophy in the Tragic Age of the Greeks*, sec. 7.
[10] Ibid., sec. 10.
[11] Ibid., sec. 13.
[12] Fr. 17, in Kirk and Raven, *The Presocratic Philosophers*, 369.
[13] KGW II/4, 331. *The Pre-Platonic Philosophers*, 123. Cf. Lange, *Geschichte des Materialismus*, 7.
[14] KGW II/4, 309. *The Pre-Platonic Philosophers*, 101.
[15] *Philosophy in the Tragic Age of the Greeks*, sec. 11.
[16] KSA 13, 14[188], 375 (*The Will to Power*, sec. 1066).
[17] See e.g. *On the Genealogy of Morals*, First Essay, sec. 13.
[18] *Philosophy in the Tragic Age of the Greeks*, sec. 13. Cf. KGW II/4, 331. *The Pre-Platonic Philosophers*, 123.
[19] *Philosophy in the Tragic Age of the Greeks*, sec. 15. This is from the *Critique of Pure Reason*, trans. Norman Kemp Smith (London: Macmillan, 1929), A37/B53, as cited in A. Spir, *Denken und Wirklichkeit: Versuch einer Erneuerung der kritischen Philosophie* (Leipzig: J. G. Findel, 1873), vol. 1, 264. A few minor alterations have occurred at both stages, not affecting the sense.
[20] *Philosophy in the Tragic Age of the Greeks*, sec. 15.
[21] Kant, *Critique of Pure Reason*, B274–79; see also B292.
[22] Spir, *Denken und Wirklichkeit*, vol. 1, 263–4. Cf. KSA 11, 35[36], 537 and KSA 11, 35[61], 538, referring to Spir, *Denken und Wirklichkeit*, vol. 2, 7 and 15.
[23] Ibid., vol. 1, 264, quoted in *Philosophy in the Tragic Age of the Greeks*, sec. 15.
[24] Spir, *Denken und Wirklichkeit*, vol. 1, 264n. Nietzsche adds just the first sentence of this passage.
[25] *Philosophy in the Tragic Age of the Greeks*, sec. 14.
[26] Plato, *Theaetetus*, 182, in *The Dialogues of Plato*, vol. 2, 185.
[27] Aristotle, *Physics* I.2. 185b, in *The Complete Works of Aristotle*, vol. 1, 317. Cf. Sextus Empiricus, *Outlines of Pyrrhonism*, I.29, 210–11, in Sextus Empiricus, *Works*, trans. R. G. Bury, Loeb Classical Library (London: Heinemann, 1933), vol. 1, 125–7.
[28] Michael S. Green, *Nietzsche and the Transcendental Tradition* (Urbana and Chicago: University of Illinois Press, 2002), 7.
[29] See e.g. KSA 9, 11[153], 500 and KSA 9, 11[330], 570; KSA 11, 26[58], 163; and KSA 12, 9[89], 382 (*The Will to Power*, sec. 517).
[30] Lange, *Geschichte des Materialismus*, 345.
[31] KSA 13, 11[73], 36 (*The Will to Power*, sec. 715).
[32] *Philosophy in the Tragic Age of the Greeks*, sec. 5.
[33] *Mixed Opinions and Maxims*, sec. 23. See also *The Wanderer and His Shadow*, sec. 11.
[34] *Twilight of the Idols*, '"Reason" in Philosophy', sec. 2.
[35] KSA 9, 11[293], 554.
[36] KSA 12, 2[151], 140 (*The Will to Power*, sec. 556). Sometimes he uses the word *Vorgang* with the same meaning, e.g. in *Beyond Good and Evil*, sec. 17.
[37] KSA 12, 2[139], 136 (*The Will to Power*, sec. 631).
[38] Spir, *Denken und Wirklichkeit*, vol. 1, 270.
[39] Ibid., vol. 1, 271.
[40] J. F. Herbart, *Lehrbuch zur Einleitung in die Philosophie*, sec. 108, in *Sämtliche Werke*, eds Karl Kehrbach and Otto Flügel (Aalen: Scientia Verlag, 1964), vol. 4, 171.
[41] Spir, *Denken und Wirklichkeit*, vol. 1, 270.

[42] KSA 13, 11[73], 36 (*The Will to Power*, sec. 708).

[43] *Twilight of the Idols*, 'How the "True World" Finally Became a Fable: The History of an Error'.

[44] KSA 11, 36[15], 556 (*The Will to Power*, sec. 1062). See also the earlier version: KSA 9, 11[292], 553.

[45] See e.g. KSA 9, 11[152], 500.

[46] See Robin Small, 'Nietzsche and Cosmology', in *A Companion to Nietzsche*, ed. Keith Ansell Pearson (Oxford: Blackwell, 2006), 189–207.

[47] KSA 13, 14[188], 375 (*The Will to Power*, sec. 1066).

[48] John Philoponus, *Against Aristotle, on the Eternity of the World*, trans. C. Wildberg (London: Duckworth, 1987).

[49] KSA 13, 14[188], 375 (*The Will to Power*, sec. 1066).

[50] Aquinas, *Summa theologiae* (London: Blackfriars, 1965–81), vol. 8, 1a q.46 art.2, 82. See also Aquinas, *De aeternitate mundi*, in *Opuscula philosophica*, ed. R. M. Spiazzi (Turin and Rome: Marietti, 1954), 103–8.

[51] KSA 9, 11[201], 522.

[52] Cicero, *De Oratore, De Fato, Paradoxa Stoicorum and De partitione oratoria*, trans. H. Rackham, Loeb Classical Library (London: Heinemann, 1940), 13.

[53] See A. N. Prior, 'Diodorean Modalities', *Philosophical Quarterly* 5 (1955): 205–13 and a considerable subsequent literature.

[54] Alexander of Aphrodisias, *Alexander of Aphrodisias on Fate*, trans. R. W. Sharples (London: Duckworth, 1983), 51.

[55] KSA 9, 11[202], 523 and KSA 13, 14[188], 376 (*The Will to Power*, sec. 1066).

[56] See e.g. Eugen Dühring, *Cursus der Philosophie als streng wissenschaftlicher Weltanschauung und Lebensgestaltung* (Leipzig: Erich Koschny, 1875), 84–5.

[57] Milič Čapek, 'The Theory of Eternal Recurrence in Modern Philosophy of Science, with Special Reference to C.S. Peirce', *Journal of Philosophy* 57 (1960): 289–96. See also Bernd Magnus, *Nietzsche's Existential Imperative* (Bloomington and London: Indiana University Press, 1978), 96–7.

[58] Hans Reichenbach, *The Direction of Time* (Berkeley, CA: University of California Press, 1956), 99, cited in Čapek, art. cit., p. 291 and *The Philosophical Impact of Contemporary Physics* (New York: Van Nostrand Reinhold Company, 1961), 127.

[59] Galileo Galilei, *On Motion and on Mechanics*, ed. and trans. I. E. Drabkin (Madison, WI: University of Wisconsin Press, 1960), 98.

[60] KSA 11, 36[15], 556 (*The Will to Power*, sec. 1062).

[61] KSA 9, 11[292], 553–4, but I have read *verwehre* for *vermehre* and *Verminderung auf dem Nichts* for *Verminderung aus dem Nichts*. Note the similar entry of four years later: KSA 11, 36[115], 556 (*The Will to Power*, sec. 1062).

[62] KSA 9, 11[201], 522.

[63] Ibid., 11[265], 543.

[64] KSA 11, 35[55], 537. Cf. KSA 11, 39[11], 622.

[65] Ibid., 35[54], 536–7 (*The Will to Power*, sec. 1064).

Chapter 2

[1] Kant, *Critique of Pure Reason*, A32/B48.

[2] KSA 11, 25[406], 118.

[3] KSA 9, 11[184], 513.

[4] John Richardson, 'Nietzsche on Time and Becoming', in *A Companion to Nietzsche*, ed. Keith Ansell Pearson (Oxford: Basil Blackwell, 2006), 210.

[5] Richardson, 'Nietzsche on Time and Becoming', 214.

[6] This sounds very like Henri Bergson: see e.g. Bergson, *Creative Evolution*, trans. Arthur Mitchell (New York: The Modern Library, 1944), 8. Keith Ansell Pearson has drawn attention to the affinities: Ansell Pearson, *Viroid Life: Perspectives on Nietzsche and the Transhuman Condition* (London and New York: Routledge, 1997), 92–3 and 'Nietzsche's Brave New World of Force: On Nietzsche's 1873 "Time Atom Theory" Fragment and the Matter of Boscovich's Influence on Nietzsche', *Journal of Nietzsche Studies*, no. 20 (2000): 12–14.

[7] Werner Stegmaier, 'Zeit der Vorstellung: Nietzsches Vorstellung der Zeit', *Zeitschrift für philosophische Forschung* 41 (1987): 225.

[8] Ibid., 227.

[9] KSA 11, 25[376], 110 and KSA 11, 40[49], 653. The likeliest source for this claim is Baudelaire, *Les Paradis artificiels*, ed. J. Crépet (Paris: Louis Conard, 1928), 34.

[10] KSA 9, 11[260], 539–40. See also KSA 12, 4[5], 178.

[11] See e.g. Cicero, *The Verrine Orations*, trans. L. H. G. Greenwood, Loeb Classical Library (London: Heinemann, 1935), II.IV 43, 398; II.V 35, 568 and II.V 72, vol. 2, 674. The next expression ἐν ἀωρονυχτι refers to the first chorus in Aeschylus's *Choephori* (*The Libation Bearers*). Finally, Homer's phrase 'time of night milking', which Nietzsche finds puzzling, is clarified in George M. Bolling, 'NYKTOΣ AMOΛΓΩI', *The American Journal of Philology*, 79.2 (1958), 165–72.

[12] *Thus Spoke Zarathustra*, 'The Drunken Song', sec. 3. Cf. ibid., sec. 8. 'Overawake' is Walter Kaufmann's translation of *überwache*.

[13] See e.g. KSA 10, 9[3], 346.

[14] Again, in the chapter 'At Noon' Zarathustra emerges from an experience of timelessness 'as from a strange drunkenness'.

[15] *The Birth of Tragedy*, sec. 1.

[16] *Thus Spoke Zarathustra*, 'On the Three Evils'.

[17] KGW II/3, 401. Cf. KGW II/3, 279.

[18] KSA 9, 6[340], 283. For other uses of *nachhinken*, see KSA 9, 11[316], 564 and *Thus Spoke Zarathustra*, 'On the Way of the Creator'.

[19] Joan Stambaugh, *The Problem of Time in Nietzsche*, trans. John F. Humphrey (Lewisburg, PA: Bucknell University Press, 1987), 67. See also Stambaugh, *Nietzsche's Thought of Eternal Return* (Baltimore and London: The Johns Hopkins Press, 1972), 114.

[20] KSA 9, 12[160], 603.

[21] KSA 7, 19[187], 477.

[22] *Human, All-Too-Human*, sec. 13. Cf. KSA 8, 21[38], 372.

[23] *Twilight of the Idols*, 'The Four Great Errors', sec. 4. Cf. KSA 11, 26[35], 156–7.

[24] KSA 11, 26[44], 159. Cf. KSA 11, 34[53], 437.

[25] KSA 9, 11[281], 549

[26] KSA 12, 1[92], 33.

[27] KSA 1, 786–7.

[28] KSA 11, 26[231], 210 (*The Will to Power*, sec. 686).

[29] KSA 13, 14[174], 360 (*The Will to Power*, sec. 702).

[30] Ibid., 14[157], 342 (*The Will to Power*, sec. 778). Cf. *Beyond Good and Evil*, sec. 244: 'The German himself *is* not, he *becomes*, he "develops" (*"entwickelt sich"*).'

[31] *Daybreak*, sec. 26. The biological concept of 'mimicry' as a favourable adaptation is due to Alfred Russel Wallace rather than Charles Darwin. Cf. *Twilight of the Idols*, 'Skirmishes of an Untimely Man', sec. 14.

[32] *Ecce Homo* is subtitled 'How One Becomes What One is', and the phrase occurs again in 'Why I Am So Clever', sec. 9. See also *The Gay Science*, sec. 270 and 335, and *Thus Spoke Zarathustra*, 'The Honey Sacrifice'.

[33] *Ecce Homo*, 'Why I Write Such Good Books', 'The Untimely Ones', sec. 3.

[34] KSA 10, 24[28], 662 (*The Will to Power*, sec. 673).

[35] *The Gay Science*, sec. 112.

[36] *Twilight of the Idols*, 'The Four Great Errors', sec. 7.

[37] *The Wanderer and His Shadow*, sec. 28.

[38] *Daybreak*, sec. 13.

[39] KSA 11, 25[158], 55.

[40] KSA 10, 24[28], 662 (*The Will to Power*, sec. 673). Cf. KSA 11, 27[71], 292 and Joan Stambaugh, 'Zufall', *Philosophy Today* 43.1 (1999): 95–9.

[41] Aristotle, *Metaphysics* VI.3, 1027b, in *The Complete Works of Aristotle*, vol. 2, 1622.

[42] See Lange, *Geschichte des Materialismus*, 8.

[43] Boscovich, *A Theory of Natural Philosophy*, trans. J. M. Child (Cambridge, MA: MIT Press, 1966), 191.

[44] KSA 9, 11[228], 529.

[45] KSA 13, 14[188], 375 (*The Will to Power*, sec. 1066). Cf. *On the Genealogy of Morals*, Second Essay, sec. 16.

[46] *On the Genealogy of Morals*, Second Essay, sec. 16.

[47] KSA 9, 4[288], 171.

[48] 'All things are entangled, ensnared, enamoured (*verkettet, verfädelt, verliebt*) . . .' *Thus Spoke Zarathustra*, 'The Drunken Song', sec. 10.

[49] KSA 10, 5[1], no. 239, 215 and KSA 10, 12[8], 401; also *Thus Spoke Zarathustra*, 'On the Vision and Riddle', sec. 2, and 'The Convalescent', sec. 2.

[50] KGW II/4, 272–3, *The Pre-Platonic Philosophers*, 65, referring to *Phaedo*, 60, in *The Dialogues of Plato*, vol. 1, 443.

[51] *Thus Spoke Zarathustra*, 'The Convalescent', sec. 2. The other use is in 'On the Vision and Riddle', sec. 2.

[52] KSA 9, 12[160], 603.

[53] KSA 12, 384 (*The Will to Power*, sec. 552).

[54] KSA 9, 11[121], 484.

[55] *The Gay Science*, sec. 109.

[56] Babette E. Babich, 'A Note on Nietzsche's "Chaos sive Natura"', *New Nietzsche Studies* 5.3/4 and 6:1/2 (2003–4): 62–3.

[57] KSA 12, 9[91], 384 (*The Will to Power*, sec. 552). Kaufmann and Hollingdale's translation substitutes 'perspective' for 'optical measure', contributing to an overestimation of Nietzsche's use of this term.

[58] KSA 11, 27[31], 283. See also KSA 12, 5[36], 197 and KSA 12, 6[14], 238 (*The Will to Power*, sec. 563 and 565). Similarly in his lectures on the pre-Platonic philosophers, he says that for Democritus 'all qualities are to be traced back to variations in quantity'. KGW II/4, 333. *The Pre-Platonic Philosophers*, 124.

[59] Gaston Bachelard, *L'intuition de l'instant* (Paris: Denoël, 1985), 105.
[60] KSA 12, 9[62], 369 (*The Will to Power*, Sec. 580).
[61] *Ecce Homo*, 'Why I Write Such Good Books', 'Thus Spoke Zarathustra', sec. 3.
[62] *Beyond Good and Evil*, sec. 28. Perhaps inconsistently, the Preface that Nietzsche added to *Daybreak* at this time defines that book's tempo as 'lento'.
[63] *Ecce Homo*, Preface, sec. 4.
[64] *Twilight of the Idols*, 'The Four Great Errors', sec. 1, and *Ecce Homo*, 'Why I Am So Wise', sec. 1.
[65] *On the Genealogy of Morals*, Third Essay, sec. 25.
[66] *Thus Spoke Zarathustra*, 'On the Preachers of Death'.
[67] KSA 7, 19[210], 484.
[68] Aristotle, *Physics*, VI.2, 232a–233b, in *The Complete Works of Aristotle*, vol. 1, 392–4.
[69] Spir, *Denken und Wirklichkeit*, vol. 2, 9.
[70] KSA 9, 11[281], 549.
[71] Plato, *Laws*, 665, in *The Dialogues of Plato*, vol. 2, 442.
[72] KSA 7, 3[19], 65 and KSA 1, 574.
[73] KGW II/3, 309.
[74] KSA 1, 557 and KSA 1, 585; see also KGW II/3, 322.
[75] KSA 1, 574–5.
[76] KGW II/3, 338 and KSA 1, 574.
[77] *The Birth of Tragedy*, sec. 2.
[78] J. P. Eckermann, *Conversations of Goethe with Eckermann*, trans. J. Oxenford (London: J. M. Dent, 1930), 303.
[79] KGW II/3, 329. For a perceptive discussion, see James I. Porter, *Nietzsche and the Philology of the Future* (Stanford, CA: Stanford University Press, 2000), 160–2.
[80] KGW II/3, 338. Cf. KSA 8, 528.
[81] KSA 7, 9[116], 318.
[82] Aristoxenus, *Elementa Rhythmica*, ed. and trans. Lionel Pearson (Oxford: Clarendon Press, 1990), 23.
[83] KGW II/3, 104.
[84] Porter, *Nietzsche and the Philology of the Future* and Friederike Felicitas Günther, *Rhythmus beim frühen Nietzsche* (Berlin and New York: Walter de Gruyter, 2008).
[85] KGW II/3, 135.
[86] Ibid., 309.
[87] Letter of mid-April 1886 to Carl Fuchs. KSB 7, 178.
[88] KSA 8, 9[1], 145, referring to Eugen Dühring, *Der Werth des Lebens. Eine philosophische Betrachtung* (Breslau: Eduard Trewendt, 1865), 41. See Günther, *Rhythmus beim frühen Nietzsche*, 150–9.
[89] KSA 11, 26[275], 222. Cf. KSA 11, 39[16], 626.
[90] KSA 8, 9[1], 146. Cf. KSA 8, 11[35], 227.
[91] *Assorted Opinions and Maxims*, sec. 134. Cf. *Nietzsche Contra Wagner*, 'Wagner as a Danger', sec. 1.
[92] Wagner, *Gesammelte Schriften und Dichtungen*, ed. Wolfgang Golther (Leipzig: E. W. Fritzsch, 1871–83; reprint, Berlin: Deutsches Verlagshaus Bong & Co., 1913), vol. 9, 80.
[93] Ibid., vol. 7, 105.
[94] Ibid., 106.
[95] Ibid., 107.

[96] *Assorted Opinions and Maxims*, sec. 134 and *Nietzsche Contra Wagner*, 'Wagner as a Danger', sec. 1.

[97] Letter of mid-April 1886 to Carl Fuchs. KSB 7, 176–7. Cf. *The Wagner Case*, sec. 7.

[98] KSA 8, 11[12], 194 and KSA 8, 11[41], 233–4.

[99] KSA 11, 26[21], 154 and KSA 13, 14[61], 246–7.

[100] *Ecce Homo*, 'Why I Write Such Good Books', sec. 4.

[101] KSA 11, 38[10], 608.

Chapter 3

[1] *Philosophy in the Tragic Age of the Greeks*, sec. 5.

[2] Schopenhauer, *Parerga and Paralipomena*, trans. E. F. J. Payne (Oxford: Clarendon Press, 1974), vol. 2, sec. 143, 283.

[3] Aristotle, *Metaphysics*, IV.5. 1010a, in *The Complete Works of Aristotle*, vol. 2, 1594–5. Nietzsche alludes to this in *On the Uses and Disadvantages of History for Life*, sec. 1.

[4] Ibid., sec. 1.

[5] KSA 7, 26[12], 579.

[6] Karl Schlechta and Anni Anders, *Friedrich Nietzsche: Von den verborgenen Anfängen seines Philosophierens* (Stuttgart-Bad Cannstatt: Friedrich Frommann Verlag, 1962), 140–53. It is mentioned briefly in Alwin Mittasch, *Friedrich Nietzsche als Naturphilosoph* (Stuttgart: Alfred Kröner Verlag, 1952), 52–3.

[7] Nietzsche was writing in an already used notebook, and some older entries appear on the same pages as the 'time-atom' text. Several lines belonging to his public lectures on 'our educational institutions' can be seen on a page reproduced in Schlechta and Anders, *Friedrich Nietzsche*, 145. See KSA 7, 8[108], 264, and KGW III/5/2, 1351 for a full explanation. My thanks to Thomas Brobjer for his assistance on this point.

[8] KSA 13, 14[121], 300–1 (*The Will to Power*, sec. 688).

[9] Helmholtz, *Wissenschaftliche Abhandlungen*, (Leipzig: Johann Ambrosius Barth, 1882), vol. 1, 13.

[10] Zöllner, *Über die Natur der Cometen. Beiträge zur Geschichte und Theorie der Erkenntnis*, 2nd edn, (Leipzig: Wilhelm Engelmann, 1872), 314.

[11] Ibid., 318.

[12] Ibid., 316.

[13] Carl Neumann, *Die Principien der Elektrodynamik. Eine mathematische Untersuchung* (Tübingen: Heinrich Laupp, 1868), 38. This is quoted by Zöllner in *Über die Natur der Cometen*, 338.

[14] Boscovich, *A Theory of Natural Philosophy*, art. 31, 27.

[15] Whitlock, 'Roger Boscovich, Benedict de Spinoza and Friedrich Nietzsche: The Untold Story', *Nietzsche-Studien* 25 (1996): 215.

[16] See e.g. KSA 13, 14[184], 371; KSA 13, 14[186], 373 and KSA 13, 14[188], 376 (*The Will to Power*, sec. 567, 636 and 1066). He also uses 'atoms of force': KSA 11, 36[20] and KSA 11, 36[22], 560 (*The Will to Power*, sec. 637 and 642), and 'points of force': KSA 11, 40[36], 636 and KSA 12, 2[69], 92.

[17] Boscovich, *A Theory of Natural Philosophy*, art. 142, 60.

[18] KSA 13, 14[79], 258 (*The Will to Power*, sec. 634).

[19] KSA 7, 21[16], 527–8.

[20] As Porter points out, this shows that Nietzsche is 'diametrically at odds' with Boscovich. Porter, 'Untimely Meditations: Nietzsche's *Zeitatomistik* in Context', *Journal of Nietzsche Studies*, no. 20 (2000): 79.

[21] *Philosophy in the Tragic Age of the Greeks*, sec. 5.

[22] Schopenhauer, *The World as Will and Representation*, trans. E. F. J. Payne (New York: Peter Smith, 1969), vol. 1, 8–9. See also ibid., vol. 2, 47 and Schopenhauer, *On the Fourfold Root of the Principle of Sufficient Reason*, trans. E. F. J. Payne (La Salle, IL: Open Court, 1974), 118–19.

[23] *Philosophy in the Tragic Age of the Greeks*, sec. 5.

[24] KSA 7, 19[210], 484.

[25] Ibid., 21[17], 528 (in Daniel Breazeale's translation).

[26] Ibid., 26[12], 575. A translation of this text by Carol Diethe (with modifications by Keith Ansell Pearson) appeared in *Journal of Nietzsche Studies* 20 (2000): 1–4. My version is guided by theirs but adopts different readings in some places.

[27] Zöllner, *Über die Natur der Cometen*, 316n.

[28] Schlechta and Anders, *Friedrich Nietzsche*, 145.

[29] Greg Whitlock, 'Examining Nietzsche's "Time-atom" Fragment from 1873', *Nietzsche-Studien* 26 (1997): 353.

[30] Ibid., 354.

[31] A third diagram, unnoticed by commentators, can be seen on a page reproduced by Schlechta and Anders in *Friedrich Nietzsche*, 147. Nietzsche intersperses his remarks about slowness of force and temporal distance with three vertical strokes, symbolizing the different intervals.

[32] In the diagram as given by Claudia Crawford, the two sequences of points are identical. Crawford, 'Nietzsche's Overhuman: Creating on the Crest of the Time-point', *Journal of Nietzsche Studies*, no. 30 (2005): 23. I take it that this is because it is supposed to picture, not the 'interrupted timelines' (as on my account), but the next stage of argument, in which a reproducing being 'retains earlier time-moments alongside present ones'.

[33] KSA 11, 40[49], 653.

[34] Schlechta and Anders read *ist* rather than *in* here, making better sense, but the reproduced page confirms the Colli and Montinari version.

[35] Galileo Galilei, *Two New Sciences, Including Centers of Gravity & Force of Percussion*, trans. Stillman Drake (Madison, WI: University of Wisconsin Press, 1974), 166.

[36] I have not translated *Beschleunigung* as 'acceleration', because here it means only that some forces are faster than others. Nietzsche does use the same word in a musical context to mean an increase in tempo: KGW II/3, 136.

[37] Whitlock, 'Examining Nietzsche's "Time-atom" Fragment from 1873', 357.

[38] See Boscovich, *A Theory of Natural Philosophy*, art. 32, 27.

[39] *Ecce Homo*, 'Why I Write Such Good Books', 'Thus Spoke Zarathustra', sec. 3.

[40] Aristoxenus, *Elementa Rhythmica*, 11. Cf. KGW II/4, 131.

[41] Herbart, *Allgemeine Metaphysik*, sec. 288, in *Sämtliche Werke*, vol. 8, 181–2.

[42] Ibid., sec. 290, 184.

[43] I read *Es ist möglich, also* on the basis of the photographic reproduction of this page in Schlechta and Anders, *Friedrich Nietzsche*, 275.

[44] Zöllner, *Über die Natur der Cometen*, 321.

[45] Ibid., 325.

[46] See e.g. KSA 7, 21[16], 527–8.

[47] Ibid., 21[16], 528.
[48] KSA 13, 14[95], 273 (*The Will to Power*, sec. 633). See also KSA 13, 14[98], 275 (*The Will to Power*, sec. 551).
[49] See e.g. Nietzsche's letter of 20 March 1882 to Heinrich Köselitz. KSB 6, 183.
[50] Johannes Gustav Vogt, *Die Kraft. Eine real-monistische Weltanschauung* (Leipzig: Haupt und Tischler, 1878).
[51] See *Human, All-Too-Human*, sec. 37, *Beyond Good and Evil*, sec. 12 and Schopenhauer, *The World as Will and Representation*, vol. 2, 160–87.
[52] See e.g. KSA 13, 14[95], 273 and KSA 13, 14[98], 275 (*The Will to Power*, sec. 633 and 551).
[53] *On the Uses and Disadvantages of History for Life*, sec. 7.
[54] *Beyond Good and Evil*, sec. 150.
[55] *The Gay Science*, sec. 60.
[56] KSA 11, 34[247], 504.
[57] KSA 10, 12[27], 404.

Chapter 4

[1] *Thus Spoke Zarathustra*, 'The Child with the Mirror'.
[2] KSA 9, 13[9], 620 and KSA 9, 17[18], 668.
[3] *Thus Spoke Zarathustra*, 'On Reading and Writing'. Cf. *Thus Spoke Zarathustra*, 'The Dancing Song'.
[4] See e.g. Tracy B. Strong, *Friedrich Nietzsche and the Politics of Transfiguration* (Berkeley, CA: University of California Press, 1975), 263; Laurence Lampert, *Nietzsche's Teaching: An Interpretation of* Thus Spoke Zarathustra (New Haven and London: Yale University Press, 1986), 162; Kathleen Marie Higgins, *Nietzsche's* Zarathustra (Philadelphia: Temple University Press, 1987), 175 and Greg Whitlock, *Returning to Sils-Maria: A Commentary to Nietzsche's 'Also sprach Zarathustra'* (New York: Peter Lang, 1990), 196. Gary Shapiro identifies the story with a vision experienced during his two days of seclusion on the ship. Shapiro, *Nietzschean Narratives* (Bloomington and Indianapolis: Indiana University Press, 1989), 75. Robert Gooding-Williams sees it as a dream, like the one described in an earlier chapter, 'The Soothsayer'. Gooding-Williams, *Zarathustra's Dionysian Modernism* (Stanford, CA: Stanford University Press, 2001), 215ff.
[5] See e.g. Hans Weichelt, *Zarathustra-Commentar*, 2nd edn. (Leipzig: Felix Meiner, 1922), 124; and Giorgio Colli and Mazzino Montinari's editorial note in KGW VI/4, 899.
[6] For expert opinions on the Hamlet murder case, see David I. Macht, 'A Physiological and Pharmacological Appreciation of *Hamlet* Act I, Scene 5, Lines 59–73', *Bulletin of the History of Medicine* 23 (1949): 186–94; and R. R. Simpson, *Shakespeare and Medicine* (Edinburgh: E. and S. Livingstone, 1959), 134–40.
[7] Greg Whitlock, *Returning to Sils-Maria*, 197. Cf. Gustav Naumann, *Zarathustra-Kommentar* (Leipzig: Verlag von H. Haeffel, 1899–1901), vol. 3, 21. Naumann describes this as 'a mediaeval torture process depicted in old pictures of hell', but Whitlock takes it to be an actual practice, adding that 'This torture is portrayed in Shakespeare's *Hamlet*.'
[8] KGW VI/4, 338 and KGW VI/4, 339.

9 Twain, *Jim Smileys berühmte Springfrosch und Dergleichen wunderliche Kauze Mehr*, trans. Moritz Busch (Leipzig: Grunow, 1874). These occasions are described in Sander L. Gilman, ed., *Conversations with Nietzsche: A Life in the Words of His Contemporaries*, trans. David J. Parent (New York: Oxford University Press, 1987), 52.

10 KSA 10, 22[1], 611.

11 KSA 14, 306. This dream probably occurred during Nietzsche's summer 1877 vacation in the Bernese Oberland, since a notebook entry of that time contains the words 'Alpa, Alpa'. (KSA 8, 23[197], 474. See also KSA 9, 10[B26], 418.) See also *Thus Spoke Zarathustra*, 'The Soothsayer'.

12 Not 'many kilometers above', as Hugh J. Silverman understandably but incorrectly claims: 'The Inscription of the Moment: Zarathustra's Gate', *International Studies in Philosophy* 24 (1992): 56.

13 *Ecce Homo*, 'Why I Write Such Good Books', 'Thus Spoke Zarathustra', sec. 4. On the other hand, given that he took a local train service from Nice to Èze-sur-mer, where the path begins, rather than walking there from his lodging, this may be true.

14 *Thus Spoke Zarathustra*, 'On Human Prudence'.

15 KSA 8, 6[48], 115–16. This translation is taken from Nietzsche, *Philosophy and Truth: Selections from Nietzsche's Notebooks of the Early 1870s*, trans. and ed. Daniel Breazeale (Atlantic Highlands, NJ: Humanities Press, 1979), 144–5.

16 Lichtenberg, *Schriften und Briefe* (Darmstadt: Wissenschaftliche Buchgesellschaft, 1972), vol. 3, 43–4.

17 Ibid., 36 and 43.

18 Plato, *Phaedrus*, 246 and 253–4, in *The Dialogues of Plato*, vol. 1, 250–1 and 257–8.

19 *Daybreak*, sec. 263. Cf. *Richard Wagner in Bayreuth*, sec. 2.

20 *Thus Spoke Zarathustra*, 'On Human Prudence'.

21 Ibid., 'Before Sunrise'.

22 *The Gay Science*, sec. 60.

23 *Thus Spoke Zarathustra*, 'On the Tarantulas'.

24 Ibid., 'On Old and New Tablets', sec. 19.

25 See *The Gay Science*, 'Joke, Cunning and Revenge', no. 60 ('Higher Men'); and Part Four of *Thus Spoke Zarathustra*.

26 See Xenophon, *Memorabilia and Oeconomicus*, trans. E. C. Marchant, Loeb Classical Library (London: Heinemann, 1923), 95–103.

27 *Schopenhauer as Educator*, sec. 6.

28 For 'steps', see e.g. *The Gay Science*, sec. 44; *Thus Spoke Zarathustra*, 'On the Tree on the Mountainside', 'On the Tarantulas' and 'The Welcome'; and *Twilight of the Idols*, 'Maxims and Arrows', sec. 42. For 'ladders', see *Schopenhauer as Educator*, sec. 1; and *Thus Spoke Zarathustra*, 'On the Thousand and One Goals' and 'On Old and New Tablets', sec. 19. The narrowness of the mountain path links it with the 'tightrope' and 'bridge' of the Prologue as well.

29 KGW II/3, 144.

30 See *The Gay Science*, 'Joke, Cunning and Revenge', no. 61, and sec. 382; *Thus Spoke Zarathustra*, 'The Stillest Hour'; and *Ecce Homo*, 'Thus Spoke Zarathustra', sec. 2.

31 K. E. von Baer, *Reden gehalten in wissenschaftlichen Versammlungen und kleinere Aufsätze vermischten Inhalts* (St Petersburg: Verlag der Kaiserlichen Hofbuchhandlung H. Schmitzdorff., 1864–70; reprint, New York: Arno Press, 1978), vol. 1, 239.

[32] Ibid., 252.
[33] Ibid., 264.
[34] Ibid., 268.
[35] KGW II/4, 267. *The Pre-Platonic Philosophers*, 60.
[36] KGW II/4, 269–70. *The Pre-Platonic Philosophers*, 62. My reading of the last sentence differs from Whitlock's.
[37] Baer describes these individuals as, respectively, a Bushman, a Hottentot, a Boer and someone 'in Cape Town' who one supposes is meant to be English, or perhaps German.
[38] Baer, *Reden gehalten in wissenschaftlichen Versammlungen*, vol. 1, 280.
[39] Ibid., vol. 1, 274.
[40] See e.g. KSA 12, 7[34], 306 and *The Gay Science*, sec. 373.
[41] Teichmüller, *Die wirkliche und die scheinbare Welt. Neue Grundlegung der Metaphysik* (Breslau: Wilhelm Koebner, 1882), 305.
[42] William James, *Principles of Psychology* (London: Macmillan, 1918), vol. 1, 609 and 613.
[43] Herbart, *Psychologie als Wissenschaft*, sec. 144, in *Sämtliche Werke*, vol. 6, 227. Spir cites this discussion in *Denken und Wirklichkeit*, vol. 2, 9.
[44] *The Gay Science*, 'Joke, Cunning and Revenge', no. 48.
[45] KSA 9, 11[203], 523. These 'periodic illnesses', placed between annual and daily cycles, are presumably what Nietzsche elsewhere calls 'the repulsive natural functions (*widerlichen Natürlichkeiten*) to which every woman is subject'. *The Gay Science*, sec. 59.
[46] KGW II/4, 273. *The Pre-Platonic Philosophers*, 65. Cf. Plutarch, *Letter of Consolation to Apollinius*, trans. F. C. Babbitt, 106; in *Moralia*, Loeb Classical Library (Cambridge, MA: Harvard University Press, 1956), vol. 2, 133.
[47] *Ecce Homo*, 'Why I Am So Wise', sec. 1.
[48] KSA 9, 11[184], 513.
[49] *Human, All-Too-Human*, sec. 18.
[50] *The Gay Science*, sec. 112.
[51] KSA 11, 40[49], 653.
[52] KSA 7, 19[174], 473.
[53] Baer, *Reden gehalten in wissenschaftlichen Versammlungen*, vol. 1, 268.
[54] *Thus Spoke Zarathustra*, 'On the Tree on the Mountainside'.
[55] KSA 9, 12[128], 598.
[56] R. W. Emerson, *The Collected Works of Ralph Waldo Emerson*, ed. Alfred R. Ferguson et al. (Cambridge, MA: The Bellknap Press of Harvard University Press, 1971–), vol. 2, 33.
[57] See *Daybreak*, sec. 182 and 301.
[58] *Beyond Good and Evil*, sec. 212.
[59] *The Gay Science*, sec. 343.
[60] *Critique of Pure Reason*, B48–49. See Spir, *Denken und Wirklichkeit*, vol. 2, 5 and 11.
[61] Schopenhauer, *On the Fourfold Root of the Principle of Sufficient Reason*, sec. 18, 46.
[62] Aristotle, *Categories*, chap. 5, 4a, trans. J. L. Ackrill, in *The Complete Works of Aristotle*, vol. 1, 7.
[63] *Thus Spoke Zarathustra*, 'On the Despisers of the Body'.
[64] KSA 12, 9[91], 385 (*The Will to Power*, sec. 552).

Chapter 5

1 Aristotle, *Rhetoric* III.2, 1405b, in *The Complete Works of Aristotle*, vol. 2, 2241.
2 Plato, *Parmenides* 152B, trans. F. M. Cornford, in Cornford, *Plato and Parmenides* (London: Routledge & Kegan Paul, 1939), 187.
3 *On the Uses and Disadvantages of History for Life*, sec. 1.
4 Fr. 1, in Kirk and Raven, *The Presocratic Philosophers*, 267.
5 Aristotle, *Physics* IV.11, 220a, in *The Complete Works of Aristotle*, vol. 1, 372.
6 Cf. St Augustine, *Confessions*, trans. R. S. Pine-Coffin (Harmondsworth: Penguin Books, 1961), X.15, 265.
7 Aristotle, *Physics* VI.3, 234a, in *The Complete Works of Aristotle*, vol. 1, 395. Cf. Teichmüller, *Die wirkliche und die scheinbare Welt*, 206.
8 It is not a 'blinking', as some commentators think. A glance is just the opposite: it is an act of perception, and this positive connotation is quite important.
9 Plato, *Timaeus*, 37–8, in *The Dialogues of Plato*, vol. 2, 19.
10 KGW II/4, 292. *The Pre-Platonic Philosophers*, 85 (translation modified).
11 KGW II/4, 309. *The Pre-Platonic Philosophers*, 101.
12 *On the Uses and Disadvantages of History for Life*, sec. 1.
13 Ibid., sec. 1.
14 KSA 9, 6[324], 280. Cf. *Daybreak*, sec. 49.
15 *Thus Spoke Zarathustra*, 'On Redemption'.
16 Ibid., 'The Soothsayer'.
17 Martin Heidegger, *Nietzsche*, trans. D. F. Krell (San Francisco: Harper and Row, 1979–87), vol. 2, 224, and *What is Called Thinking?* trans. Fred D. Wieck and J. Glenn Gray (New York: Harper and Row, 1968), 96.
18 Ofelia Schutte, 'Willing Backwards: Nietzsche on Time, Pain, Joy, and Memory', in *Nietzsche and Depth Psychology*, ed. Jacob Golomb, Weaver Santaniello and Ronald L. Lehrer (Albany, NY: State University of New York Press, 1999), 122.
19 KSA 13, 14[188], 375 (*The Will to Power*, sec. 1066).
20 Joan Stambaugh, 'Das Gleiche in Nietzsches Gedanken der Ewigen Wiederkunft des Gleichen', *Revue internationale de philosophie* 18, no. 67 (1964): 95. My translation.
21 KSA 10, 4[88], 140 and KSA 10, 5[1], no. 147, 203.
22 J. M. E. McTaggart, 'The Unreality of Time', *Mind* 17 (1908), 457–74, and *The Nature of Existence* (Cambridge: Cambridge University Press, 1938; reprint, New York: Octagon Books, 1976), vol. 2, chap. XXXIII.
23 McTaggart, 'The Unreality of Time', 458 and *The Nature of Existence*, vol. 2, 10.
24 McTaggart, *The Nature of Existence*, vol. 2, 10–11.
25 Schopenhauer, *The World as Will and Representation*, vol. 1, sec. 54, 278.
26 In contrast, Robert Gooding-Williams sees the gateway not only as knotted to future events but as a 'mobile entity' that itself moves along the lanes. Gooding-Williams, *Zarathustra's Dionysian Modernism*, 218. I have tried to picture this, but without success.
27 Similarly, 'All things are entangled, ensnared, enamoured . . .' *Thus Spoke Zarathustra*, 'The Drunken Song', sec. 10.
28 Ibid., 'On Redemption'.
29 KSA 10, 11[5], 380.
30 Plato, *Parmenides*, 152, in *The Dialogues of Plato*, vol. 2, 120.

[31] KSA 9, 11[281], 549.
[32] Eugen Fink, *Nietzsche's Philosophy*, trans. Goetz Richter (London and New York: Continuum, 2002), 76.
[33] Crawford, 'Nietzsche's Overhuman: Creating on the Crest of the Timepoint': 39. Similarly, Alistair Moles writes, 'Time in Nietzsche's cosmos is nothing else than the totality of recurrent moments.' *Nietzsche's Philosophy of Nature and Cosmology* (New York: Peter Lang, 1990), 295.
[34] Other suggestions are that he arrives at a crossroad, or at the top of the mountain path: see respectively, Kathleen M. Higgins, *Nietzsche's Zarathustra*, 175 and Gary Shapiro, *Nietzschean Narratives*, 75.
[35] Its converse may be occurring in the return to a mountain setting at the end of the chapter, although no path is mentioned there.
[36] Aristotle, *Physics* IV.11, 220a, in *The Complete Works of Aristotle*, vol. 1, 373; see also Aristotle, *Physics* VIII.8, 262a–b, *The Complete Works of Aristotle*, vol. 1, 437–8.
[37] Gooding-Williams, *Zarathustra's Dionysian Modernism*, 219 and 221.
[38] Ibid., 219.
[39] Stanley Rosen, *The Mask of Enlightenment: Nietzsche's Zarathustra* (Cambridge and New York: Cambridge University Press, 1995), 187.
[40] Loeb, 'The Gateway-Augenblick', in *Nietzsche's* Thus Spoke Zarathustra: *Before Sunrise*, ed. James Luchte (London and New York: Continuum, 2008), 101–2.
[41] Rosen, *The Mask of Enlightenment*, 183.
[42] Ibid., 184.
[43] KSA 13, 302 (*The Will to Power*, sec. 625).
[44] *Beyond Good and Evil*, sec. 129.
[45] Hannah Arendt, *The Life of the Mind* (London: Secker and Warburg, 1978), vol. 1, 205.
[46] Kafka, *The Great Wall of China*, trans. W. Muir and E. Muir (New York: Schocken Books, 1946), 276.
[47] Arendt, *The Life of the Mind*, vol. 1, 204.
[48] Ibid., 208.
[49] Ibid., 210 and 211.
[50] Visser, 'Der unendlich kleine Augenblick', *Nietzsche-Studien* 27 (1998): 89.

Chapter 6

[1] *The Antichrist*, sec. 1. But cf. *Beyond Good and Evil*, secs 207, 208 and 224.
[2] Teichmüller, *Die wirkliche und die scheinbare Welt*, 239.
[3] Gustav Teichmüller, *Darwinismus und Philosophie* (Dorpat: C. Mattiesen, 1877), 39.
[4] Ibid., 41; see also *Die wirkliche und die scheinbare Welt*, 216 and 239.
[5] Teichmüller, *Darwinismus und Philosophie*, 43.
[6] Teichmüller, *Die wirkliche und die scheinbare Welt*, 227.
[7] Teichmüller, *Darwinismus und Philosophie*, 49.
[8] Ibid., 43.
[9] *Thus Spoke Zarathustra*, 'Upon the Blessed Isles'. Cf. KSA 10, 12[6], 401 and KSA 10, 13[17], 465. Cf.KSA 10, 4[160], 160 and KSA 10, 5[1], no. 247, 216. This expression may allude to Isaiah 40:3.
[10] *Thus Spoke Zarathustra*, Prologue, sec. 5.

[11] Ibid., 'On the Higher Man', sec. 5.
[12] Ibid., sec. 10.
[13] Paul S. Loeb, 'The Thought-Drama of Eternal Recurrence', *The Journal of Nietzsche Studies*, no. 34 (2007): 85.
[14] *Thus Spoke Zarathustra*, 'On Reading and Writing'.
[15] Ibid., 'On the Spirit of Gravity', sec. 2.
[16] Lampert, *Zarathustra's Teaching*, 162.
[17] Heidegger, *Nietzsche*, vol. 2, 57.
[18] *Human, All-Too-Human*, sec. 18, KSA 7, 10[1], 342, and KSA 9, 4[36], 109. See Debra B. Bergoffen, 'On Nietzsche's Moles', in *A Nietzschean Bestiary: Becoming Animal Beyond Docile and Brutal*, ed. Christa Davis Acampora and Ralph R. Acampora (New York: Rowman & Littlefield, 2003).
[19] *Daybreak*, sec. 271 What Nietzsche actually wrote was probably 'the blind law of gravity'. His wording was amended during transcription by Heinrich Köselitz to fit Robert Mayer's concept of force: see Small, *Nietzsche in Context* (Aldershot: Ashgate, 2001), 140–1.
[20] *Thus Spoke Zarathustra*, 'The Convalescent', sec. 1.
[21] For a discussion of Nietzsche's use of this concept, see Moles, *Nietzsche's Philosophy of Nature and Cosmology*, 274–83.
[22] Friedrich Ulfers and Mark Daniel Cohen, 'The Energeticist Model of the Universe as *Perpetuum Mobile* and Nietzsche's Notion of the Eternal Recurrence of the Same', http://homepage.mac.com/cohenmd1/.Public/Philosophy/THE%20ENERGETICIST%20MODEL.pdf. Accessed 23 February 2009.
[23] Milič Čapek, 'The Theory of Eternal Recurrence in Modern Philosophy of Science, with Special Reference to C. S. Peirce', 295, and *The Philosophical Impact of Contemporary Physics*, 344. Elsewhere Čapek adopts a sterner tone, accusing these writers of 'surreptitiously' introducing an irreversible time. 'Eternal Return', in *The Encyclopedia of Philosophy*, ed. Paul Edwards (New York: The Macmillan Company and The Free Press, 1967), vol. 3, 63. This charge is repeated by Bernd Magnus in *Nietzsche's Existential Imperative*, 67.
[24] M. Čapek, *The New Aspects of Time: Its Continuities and Novelties* (Dordrecht: Kluwer Academic Publishers, 1991), 271.
[25] George N. Schlesinger, *Timely Topics* (Basingstoke: St Martin's Press, 1994), 131.
[26] Paul S. Loeb, 'Identity and Eternal Recurrence', in *A Companion to Nietzsche*, ed. Keith Ansell Pearson (Oxford: Basil Blackwell, 2006), 181–2.
[27] Georg Cantor, *Gesammelte Abhandlungen*, ed. E. Zermelo (1932; reprint, Hildesheim: Georg Olms, 1962), 392–3. Cf. Georg Cantor, *Gesammelte Abhandlungen*, 410–11.
[28] Kant, *Critique of Pure Reason*, A759/B787.
[29] Ibid., A762/B790.
[30] T. K. Seung, *Nietzsche's Epic of the Soul: Thus Spoke Zarathustra* (Lanham, MD: Lexington Books, 2005), 123.
[31] One of the informal fallacies of argument recognized by the ancient Nyāya school of India is 'evasion', which 'arises if one stops an argument on the pretext of going away to attend another business'. Sarvepalli Radhakrishnan and Charles A. Moore (eds), *A Source Book in Indian Philosophy* (Princeton, NJ: Princeton University Press, 1957), 378.
[32] KSA 9, 11[245], 534.

33 Bernd Magnus, *Nietzsche's Existential Imperative*, 82. See also Linda L. Williams and Joseph T. Palencik, 'Re-evaluating Nietzsche's Cosmology of Eternal Recurrence', *The Southern Journal of Philosophy* 42 (2004): 402.

34 The 'inversion of time' discussed in Chapter 2 is not an exception, since it refers only to the construction of causalities within particular experiences.

35 Lampert, *Zarathustra's Teaching*, 165.

36 See e.g. Donald Davidson, 'Events as Particulars', *Noûs* 4 (1970): 25–32.

37 W. V. O. Quine, 'Events and Reification', in *Actions and Events: Perspectives on the Philosophy of Donald Davidson*, ed. Ernest LePore and Brian P. McLaughlin (Oxford: Basil Blackwell, 1985), 162–71.

38 See e.g. Bernd Magnus, *Nietzsche's Existential Imperative*, 107–8, and – omitting quite a few in between – Robert Gooding-Williams, *Zarathustra's Dionysian Modernism*, 218.

39 R. Chisholm, 'Events and Propositions', *Noûs* 4 (1970): 15.

40 Magnus, *Nietzsche's Existential Imperative*, 98–110.

41 Gooding-Williams, *Zarathustra's Dionysian Modernism*, 218.

42 Jorge Luis Borges, 'Tlön, Uqbar, Orbius Tertius', trans. James E. Irby, in Borges, *Labyrinths: Selected Stories and Other Writings*, ed. Donald A. Yates and James E. Irby (Harmondsworth: Penguin Books, 1970), 33.

43 *Thus Spoke Zarathustra*, 'On the Vision and Riddle' and 'The Convalescent', also KSA 9, 11[206], 524; KSA 10, 24[7], 646 (*The Will to Power*, sec. 1058); and KSA 12, 7[54], 312 (*The Will to Power*, sec. 617).

44 KSA 9, 11[141], 494; KSA 9, 11[245], 534; and KSA 10, 5[1], no. 160, 205. See also KSA 13, 16[32], 492 ('the same things').

45 KSA 9, 11[202], 523; KSA 9, 11[232], 530; KSA 9, 11[292], 553; and KSA 9, 11[305], 558–9; also KSA 11, 36[15], 556.

46 KSA 9, 11[148], 498 and KSA 9, 11[245], 534; also KSA 10, 1[27], 15.

47 KSA 9, 11[202], 523 and KSA 9, 13; KSA 9, 14[188], 376 (*The Will to Power*, sec. 1066).

48 Ibid., 11[245], 534 and Ibid., 11[269], 544; also KSA 13, 14[188], 376 (*The Will to Power*, sec. 1066).

49 KSA 9, 11[292], 553 and KSA 11, 36[15], 556; and KSA 11, 667 (*The Will to Power*, sec. 1067).

50 KSA 9, 11[305], 559.

51 *Thus Spoke Zarathustra*, 'The Convalescent' and 'On the Vision and Riddle'.

52 *The Gay Science*, sec. 341, KSA 9, 11[148], 498 and KSA 9, 11[206], 524.

53 Joan Stambaugh, *Nietzsche's Thought of Eternal Return*, 30.

54 See e.g. *Nietzsche's Thought of Eternal Return*, 30 and Michael Skowron, 'Zarathustra-Lehren Übermensch, Wille zur Macht, Ewige Wiederkunft', *Nietzsche-Studien* 33 (2004): 87.

55 *The Gay Science*, sec. 341, *Thus Spoke Zarathustra*, 'On the Vision and Riddle', KSA 9, 11[148], 498 and KSA 10, 15[3], 479.

Chapter 7

1 Loeb, 'Identity and Eternal Recurrence', 185. See also Loeb, 'Time, Power, and Superhumanity', *Journal of Nietzsche Studies*, no. 21 (2001), 34.

2 Loeb, 'Identity and Eternal Recurrence', 173.

3 H. Bergson, *Mind-Energy: Lectures & Essays*, trans. H. W. Carr (London: Macmillan, 1920), 109.

4 Ibid., 136. Cf. ibid., 143.

5 Bergson's earlier theory of dreaming worked along similar lines. See Bergson, *Matter and Memory*, trans. N. M. Paul and W. S. Palmer (London: George Allen and Unwin, 1911).

6 Pierre Klossowski, *Nietzsche and the Vicious Circle*, 57.

7 Ibid., 58.

8 Rosen, *The Mask of Enlightenment*, 188.

9 Georg Simmel, *Schopenhauer and Nietzsche*, trans. Helmut Loiskandl, Deena Weinstein and Michael Weinstein (Amherst, MA: The University of Massachusetts Press, 1986), 173. Similarly, William James writes, 'it is logically impossible that the same thing should be known as the same by two successive copies of the same thought'. James, *The Principles of Psychology*, vol. 1, 480.

10 See Philip J. Kain, 'Nietzsche, Skepticism and Eternal Recurrence', *Canadian Journal of Philosophy* 13 (1983), 376–7.

11 KSA 9, 12[15], 578. Cf. KSA 10, 12[31], 406.

12 KSA 11, 25[403], 117. See Gregory Moore, *Nietzsche, Biology and Metaphor* (Cambridge: Cambridge University Press, 2002), 41.

13 This is also Bergson's reason for claiming that 'consciousness cannot go through the same state twice'. Bergson, *Creative Evolution*, 8.

14 See e.g. *The Wanderer and his Shadow*, sec. 40, a close paraphrase of Paul Rée's account of the moral sense.

15 KSA 9, 11[166], 505.

16 *Human, All-Too-Human*, sec. 19.

17 KSA 9, 11[202], 523 and KSA 9, 11[231], 530.

18 A counterpart in notebook sketches of the doctrine is the principle, discussed earlier, that 'whatever state this world merely *can* attain, it must have attained'. KSA 9, 11[148], 498.

19 *The Gay Science*, sec. 341.

20 *Beyond Good and Evil*, sec. 68.

21 Bertrand Russell, *The Problems of Philosophy* (London: Williams and Norgate, 1912), 180.

22 Williams and Palencik, 'Re-evaluating Nietzsche's Cosmology of Eternal Recurrence', 395. Cf. ibid., 408.

23 Moles, *Nietzsche's Philosophy of Nature and Cosmology*, 314.

24 Gooding-Williams, *Zarathustra's Dionysian Modernism*, 255.

25 Seung, *Nietzsche's Epic of the Soul*, 325.

26 *Thus Spoke Zarathustra*, 'On Redemption'.

27 A. N. Prior, 'Thank Goodness That's Over', *Philosophy* 34.1 (1959), 17.

28 *The Gay Science*, sec. 341. I have not followed Walter Kaufmann's replacement of 'and' with 'or' in this passage.

29 Ivan Soll, 'Reflections on Recurrence: A Re-examination of Nietzsche's Doctrine, die Ewige Wiederkehr des Gleichen', in *Nietzsche: A Collection of Critical Essays*, ed. Robert Solomon (New York: Doubleday Anchor, 1973), 339.

30 Simmel, *Schopenhauer and Nietzsche*, 173–4.

31 Henri Bois, Le "retour éternel" de Nietzsche', *L'Année philosophique* 24 (1913), 170–1. The words 'resembles me like a brother' are from Alfred de Musset's poem 'Le poète', where they refer to an *alter ego* who appears to the narrator at times of misfortune throughout his life.
32 Horace, *The Odes and Epodes*, trans. C. E. Bennett, Loeb Classical Library (London: Heinemann, 1960), III.3, 7–8, 179.
33 Maudemarie Clark, *Nietzsche on Truth and Philosophy* (Cambridge: Cambridge University Press, 1990), 268.
34 KSA 9, 11[160], 503.
35 Ibid., 11[163], 505.
36 Joan Stambaugh, 'Thoughts on a Nachlass Fragment from Nietzsche', *Nietzsche-Studien* 6 (1977), 195.
37 Revelation 22:11.
38 Revelation 10:6.
39 KSA 9, 11[163], 504–5.
40 *The Gay Science*, sec. 125.
41 *Thus Spoke Zarathustra*, Prologue, sec. 7.
42 Ibid., 'On the Afterworldly'.
43 KSA 9, 11[172], 507.
44 *Human, All-Too-Human*, sec. 220.
45 *The Gay Science*, sec. 78.
46 See e.g. KSA 13, 14[173–4], 358–62 (*The Will to Power*, secs. 699, 702, 652 and 703).
47 *Thus Spoke Zarathustra*, 'The Soothsayer'.
48 KSA 11, 26[283], 225 (*The Will to Power*, sec. 1060).
49 See e.g. KSA 10, 24[4], 645 and KSA 11, 26[376], 250 (*The Will to Power*, sec. 1057 and 1053).
50 *Daybreak*, sec. 77. See W. E. H. Lecky, *A History of England in the Eighteenth Century*, (London: Longmans, Green, and Co., 1913), vol. 3, 78–83, and Robin Small, 'Disturbing Thoughts and Eternal Perspectives: Some Uses of Symbolism in Nietzsche', *New Nietzsche Studies* 3/3-4 (1999): 30–1.
51 KSA 9, 11[159], 503.
52 Rosen, *The Mask of Enlightenment*, 15.
53 Plato, *Timaeus*, 37, in *The Dialogues of Plato*, vol. 2, 19
54 KSA 9, 11[161], 503.
55 KSA 10, 24[7], 646 (*The Will to Power*, sec. 1058).
56 See e.g. Richardson, *Nietzsche's New Darwinism*, 37–8.
57 See e.g. Hans Weichelt, *Zarathustra-Commentar*, 2nd edn, 124; and any number of later writers.
58 See e.g. *Ecce Homo*, 'Why I Am So Clever', sec. 6.
59 *Thus Spoke Zarathustra*, 'On Involuntary Bliss'.
60 One might however argue for some link between the description of the sailors as 'glad of the twilight' (*Zwielicht*) and the location of the gateway scene in 'the deadly pallor of twilight' (*Dämmerung*). Alan White argues that these words have different connotations; see White, *Nietzsche's Labyrinth* (New York and London: Routledge, 1990), 166. Still, in the later chapter 'On the Great Longing' Zarathustra speaks of 'dust, spiders and twilight' (*Zwielicht*), apparently referring to his encounter with the spirit of gravity.

[61] A. J. Arberry, *Tales From the Masnavi* (London: George Allen and Unwin, 1961), 135–6.

[62] *Daybreak*, sec. 77. See also Small, 'Disturbing Thoughts and Eternal Perspectives', 29–44.

[63] See Lecky, *A History of England in the Eighteenth Century*, vol. 3, 82–3.

Chapter 8

[1] The phrase 'that which has hitherto been your ultimate danger has become your ultimate refuge' is another allusion to Lichtenberg's Cook anecdote; and in Part Four of *Thus Spoke Zarathustra*, the 'murderer of God' tells Zarathustra: 'Now *you* are my last refuge.'

[2] See e.g. Lampert, *Nietzsche's Teaching*, 159, and Seung, *Nietzsche's Epic of the Soul*, 120 and 127.

[3] Emerson, *The Collected Works of Ralph Waldo Emerson*, vol. 2, 39.

[4] *Schopenhauer as Educator*, sec. l. Similarly, 'For he who proceeds on this own way encounters no one: that is implied in "own way".' *Daybreak*, Preface, sec. 2.

[5] *Thus Spoke Zarathustra*, 'On the Spirit of Gravity', sec. 2.

[6] The translation of *Gasse* as 'lane' may suggest a narrow path or alley, but in some German-speaking cities (such as Basel, where Nietzsche had lived for some years) main streets may be *Gassen*.

[7] *Thus Spoke Zarathustra*., 'On Virtue that Makes Small', sec. 2. See also KSA 10, 13[3], 454; KSA 10, 16[7], 500 and KSA 10, 22[1], 606.

[8] *Thus Spoke Zarathustra*, 'On the Higher Man', sec. 17. A draft reads: "Great men and currents go crookedly to their goal: crookedly, but *to their* goal. That is their best courage, that they are not afraid of crookedness." KSA 10, 22[6], 633. In the last sentence, I follow Haase and Montinari's substitution of *Muth* for *Werth*, and also venture a substitution of *nicht* for *noch*.

[9] *Thus Spoke Zarathustra*, 'The Seven Seals', sec. 7. Cf. KSA 10, 3[1], no. 256, 83; and *Beyond Good and Evil*, sec. 71.

[10] *Thus Spoke Zarathustra*, 'Before Sunrise'.

[11] *Beyond Good and Evil*, sec. 193.

[12] *Human, All-Too-Human*, sec. 611.

[13] Fr. 60, in Kirk and Raven, *The Presocratic Philosophers*, 189.

[14] *Thus Spoke Zarathustra*, 'The Voluntary Beggar'.

[15] *Ecce Homo*, 'Why I Write Such Good Books', 'Thus Spoke Zarathustra', sec. 6.

[16] *The Antichrist*, sec. 1, quoting lines 29–30 of Pindar's Tenth Pythian Ode.

[17] *Human, All-Too-Human*, sec. 15.

[18] *The Gay Science*, sec. 125.

[19] Ibid., sec. 124.

[20] Ibid., sec. 343.

[21] Ibid., sec. 374.

[22] *Twilight of the Idols*, 'The Four Great Errors', sec. 4. Cf. KSA 11, 26[44], 159 and KSA 11, 34[54], 437.

[23] *Thus Spoke Zarathustra*, 'On the Higher Man', sec. 17.

[24] Ibid., 'On the Vision and Riddle', sec. 1.

25 Ibid., 'On the Despisers of the Body'.
26 See, respectively: Prologue; 'The Stillest Hour' and 'Involuntary Bliss'; 'On Involuntary Bliss' (for *seligen Stunde*) and 'On Virtue That Makes Small', sec. 3 (for *gesegnete Stunde*).
27 *On the Genealogy of Morals*, First Essay, sec. 12 and Third Essay, sec. 7. Walter Kaufmann's translation of *Stunde* as 'moment' in the second passage contributes to the confusion of 'hour' and 'moment' found in many commentaries on *Thus Spoke Zarathustra.*
28 For 'the hour of noon' (*die Stunde des Mittags*) see 'On Great Events' and 'At Noon'; also KSA 9, 11[148], 498.
29 *Thus Spoke Zarathustra*, 'On Passing By'.
30 Ibid., 'On Old and New Tablets', sec. 1, and 'The Sign'.
31 Ibid., 'The Sign'.
32 Charles Baudelaire, *Paris Spleen*, trans. Louise Varèse (New York: New Directions, 1970), 30.
33 *Thus Spoke Zarathustra*, 'On Involuntary Bliss'.
34 Ibid., 'On Free Death'.
35 Mark 13:33 and John 7:8.
36 See Stegmaier, 'Zeit der Vorstellung', 206.
37 *The Gay Science*, sec. 377.
38 *Beyond Good and Evil*, sec. 212.
39 Stegmaier, 'Zeit der Vorstellung', 209.
40 Emerson, *The Collected Works of Ralph Waldo Emerson*, vol. 2, 38.
41 *Thus Spoke Zarathustra*, 'On the Great Longing'. Kaufmann aptly gives 'dwell closer together'.
42 Emerson, *The Collected Works of Ralph Waldo Emerson*, vol. 6, 13. Nietzsche invokes this idea of inner-outer equilibrium when discussing fatalism, but overlooks it in claiming the benefit of lower atmospheric pressure at an altitude of 6000 feet: 'Two pounds less to carry than down at sea level.' KSA 9, 11[238], 532.
43 Arendt, *The Life of the Mind*, 211.
44 Ibid., 210.
45 *On the Uses and Disadvantages of History for Life*, sec. 1. In draft versions he wrote 'between the two gates (*Thoren*) of past and future', KSA 7, 29[98], 677 and KSA 7, 30[2], 726.
46 *Ecce Homo*, 'Why I Am So Wise', sec. 3.
47 *The Gay Science*, sec. 341 and *Thus Spoke Zarathustra*, 'On the Vision and Riddle'.
48 See e.g. Higgins, *Nietzsche's* Zarathustra, 174–7 and Gooding-Williams, *Zarathustra's Dionysian Modernism*, 218–19.
49 KSA 9, 11[318], 564.
50 Ibid., 11[281], 549. This passage has been discussed in Chapter 2.
51 *Physics*, IV.11, 218b, in *The Complete Works of Aristotle*, vol. 1, 371. Cf. Spir, *Denken und Wirklichkeit*, vol. 2, 8. Nietzsche had recently retrieved his copy of Spir's book from storage at the Overbeck home when he made this notebook entry: see his letter of 20/21 August 1881, KSB 6, 118.
52 Loeb, 'The Gateway-Augenblick', 104.
53 *Thus Spoke Zarathustra*, 'The Convalescent', sec. 2.
54 Nietzsche may well have in mind Spir's statement that a time without succession is not a time at all (*ist gar keine Zeit*). Spir, *Denken und Wirklichkeit*, vol. 2, 10.

[55] I set aside the places where it means only that time has not been allowed for something: e.g. *Daybreak*, sec. 162; *The Gay Science*, sec. 32; and *Beyond Good and Evil*, sec. 58.

[56] KSA 9, 11[260], 540.

[57] *Twilight of the Idols*, 'The Four Great Errors', sec. 4.

[58] KSA 9, 11[183], 513.

[59] *Thus Spoke Zarathustra*, 'The Sign'. Note also an earlier passage: '"We have no time yet for Zarathustra," they argue; but what matters a time that "has no time" for Zarathustra?' *Thus Spoke Zarathustra*, 'On Virtue That Makes Small'.

[60] Franz Rosenzweig, *The Star of Redemption*, trans. W. H. Hallo (London: Routledge and Kegan Paul, 1971), 290.

[61] Ofelia Schutte, 'Willing Backwards: Nietzsche on Time, Pain, Joy, and Memory', 122.

[62] This 'living in the future' is not to be confused with the notion of *entering* the future discussed and rejected in Chapter 5 as a misunderstanding of the notion of a general movement of time.

[63] *On the Uses and Disadvantages of History For Life*, sec. 1.

[64] Rosenzweig, *The Star of Redemption*, 290.

[65] KSA 11, 38[12], 611 (*The Will to Power*, sec. 1067).

[66] *Gorgias*, 507–8, in *The Dialogues of Plato*, vol. 1, 569.

[67] *Beyond Good and Evil*, sec. 150, KSA 11, 38[12], 610–11 (*The Will to Power*, sec. 1067) and KSA 12, 10[90], 508 (*The Will to Power*, sec. 1037).

[68] *Thus Spoke Zarathustra*, 'On Old and New Tablets', sec. 2. See also KSA 11, 38[12], 610–11 (*The Will to Power*, sec. 1067).

[69] *The Gay Science*, sec. 85.

[70] KSA 12, 10[138], 535 (*The Will to Power*, sec. 639). Note the link with the immediately preceding note, included in *The Will to Power* as sec. 707.

[71] KSA 12, 10[90], 507–8 (*The Will to Power*, sec. 1037).

[72] Ibid., 9[8], 343 (*The Will to Power*, sec. 712).

[73] *Twilight of the Idols*, 'The Four Great Errors', sec. 2.

[74] KSA 12, 2[130], 129 (*The Will to Power*, sec. 797).

[75] Lucian, *Works*, Loeb Classical Library, vol. 2, 476.

[76] *Philosophy in the Tragic Age of the Greeks*, sec. 7. KSA 1, 830.

[77] *Thus Spoke Zarathustra*, 'The Drunken Song', sec. 10.

Bibliography

Alexander of Aphrodisias. *Alexander of Aphrodisias on Fate.* Translated by R. W. Sharples. London: Duckworth, 1983.

Ansell Pearson, Keith. 'Nietzsche's Brave New World of Force: On Nietzsche's 1873 "Time Atom Theory" Fragment and the Matter of Boscovich's Influence on Nietzsche'. *Journal of Nietzsche Studies* no. 20 (2000): 5–33.

—. *Viroid Life: Perspectives on Nietzsche and the Transhuman Condition.* London and New York: Routledge, 1997.

Aquinas, St Thomas. *Opuscula philosophica.* Edited by R. M. Spiazzi. Turin and Rome: Marietti, 1954.

—. *Summa Theologiae.* 61 vols. London: Blackfriars, 1965–81.

Arberry, A. J. *Tales From the Masnavi.* London: George Allen and Unwin, 1961.

Arendt, Hannah. *The Life of the Mind.* 2 vols. London: Secker and Warburg, 1978.

Aristotle. *The Complete Works of Aristotle.* Edited by J. Barnes. 2 vols. Princeton, NJ: Princeton University Press, 1984.

Aristoxenus. *Elementa Rhythmica.* Edited and translated by Lionel Pearson. Oxford: Clarendon Press, 1990.

Augustine, St. *Confessions.* Translated by R. S. Pine-Coffin. Harmondsworth: Penguin Books, 1961.

Babich, Babette E. 'A Note on Nietzsche's "Chaos sive Natura"'. *New Nietzsche Studies* 5.3/4 and 6:1/2 (2003–4): 48–72.

Bachelard, Gaston. *L'intuition de l'instant.* Paris: Denoël, 1985.

Baer, Karl Ernst von. *Reden gehalten in wissenschaftlichen Versammlungen und kleinere Aufsätze vermischten Inhalts.* 3 vols. St Petersburg: Verlag der Kaiserlichen Hofbuchhandlung H. Schmitzdorff, 1864–70. Reprint, New York: Arno Press, 1978.

Baudelaire, Charles. *Les Paradis artificiels.* Edited by J. Crépet. Paris: Louis Conard, 1928.

—. *Paris Spleen.* Translated by Louise Varèse. New York: New Directions, 1970.

Bergoffen, Debra B. 'On Nietzsche's Moles'. In *A Nietzschean Bestiary: Becoming Animal Beyond Docile and Brutal,* edited by Christa Davis Acampora and Ralph R. Acampora, 243–50. New York: Rowman & Littlefield, 2003.

Bergson, Henri. *Creative Evolution.* Translated by Arthur Mitchell. New York: The Modern Library, 1944.

—. *Matter and Memory.* Translated by N. M. Paul and W. S. Palmer. London: George Allen and Unwin, 1911.

—. *Mind-Energy: Lectures & Essays.* Translated by H. W. Carr. London: Macmillan, 1920.

Bois, Henri, 'Le "Retour éternel" de Nietzsche'. *L'Année philosophique* 24 (1913): 145–84.

Bolling, George M. 'ΝΥΚΤΟΣ ΑΜΟΛΓΩΙ'. *The American Journal of Philology* 79.2 (1958): 165–72.

Borges, Jorge Luis. *Labyrinths: Selected Stories and Other Writings.* Edited by Donald A. Yates and James E. Irby. Harmondsworth: Penguin Books, 1970.

Boscovich, R. J. *A Theory of Natural Philosophy.* Translated by J. M. Child. Cambridge, MA: MIT Press, 1966.

Cantor, Georg. *Gesammelte Abhandlungen.* Edited by E. Zermelo. 1932. Reprint, Hildesheim: Georg Olms, 1962.

Čapek, Milič. 'Eternal Return'. In *The Encyclopedia of Philosophy*, edited by Paul Edwards, vol. 3, 61–3. New York: The Macmillan Company and The Free Press, 1967.

—. *The New Aspects of Time: Its Continuities and Novelties.* Dordrecht: Kluwer Academic Publishers, 1991.

—. *The Philosophical Impact of Contemporary Physics.* New York: Van Nostrand Reinhold Company, 1961.

—. 'The Theory of Eternal Recurrence in Modern Philosophy of Science, with Special Reference to C. S. Peirce'. *Journal of Philosophy* 57 (1960): 289–96.

Chisholm, R. 'Events and Propositions'. *Noûs* 4 (1970): 15–24.

Cicero. *De Oratore, De Fato, Paradoxa Stoicorum and De partitione oratoria.* Translated by H. Rackham. 2 vols. Loeb Classical Library. London: Heinemann, 1940.

—. *The Verrine Orations.* Translated by L. H. G. Greenwood. 2 vols. Loeb Classical Library. London: Heinemann, 1935.

Clark, Maudemarie *Nietzsche on Truth and Philosophy* (Cambridge: Cambridge University Press, 1990), 268.

Cornford, F. M. *Plato and Parmenides.* London: Routledge & Kegan Paul, 1939.

Crawford, Claudia. 'Nietzsche's Overhuman: Creating on the Crest of the Time-point'. *Journal of Nietzsche Studies*, no. 30 (2005): 22–48.

Davidson, Donald. 'Events as Particulars', *Noûs* 4 (1970): 25–32.

Dühring, Eugen. *Cursus der Philosophie als streng wissenschaftlicher Weltanschauung und Lebensgestaltung.* Leipzig: Erich Koschny, 1875.

—. *Der Werth des Lebens. Eine philosophische Betrachtung.* Breslau: Eduard Trewendt, 1865.

Eckermann, J. P. *Conversations of Goethe with Eckermann.* Translated by J. Oxenford. London: J. M. Dent, 1930.

Emerson, R. W. *The Collected Works of Ralph Waldo Emerson.* Edited by Alfred R. Ferguson et al. Cambridge, MA: The Bellknap Press of Harvard University Press, 1971–.

Fink, Eugen. *Nietzsche's Philosophy.* Translated by Goetz Richter. London and New York: Continuum, 2002.

Galilei, Galileo, *On Motion and on Mechanics.* Edited and translated by I. E. Drabkin. Madison, WI: University of Wisconsin Press, 1960.

—. *Two New Sciences, Including Centers of Gravity & Force of Percussion.* Translated by Stillman Drake. Madison, WI: University of Wisconsin Press, 1974.

Gerhardt, Volker. 'Die Metaphysik des Werdens: Über ein traditionelles Element in Nietzsches Lehre vom "Willen zur Macht"'. In *Nietzsche und die philosophische Tradition*, edited by Josef Simon, Band I, 9–33. Würzburg: Königshausen + Neumann, 1985.

Gilman, Sander L., ed. *Conversations with Nietzsche: A Life in the Words of His Contemporaries.* Translated by David J. Parent. New York: Oxford University Press, 1987.

Gooding-Williams, Robert. *Zarathustra's Dionysian Modernism.* Stanford, CA: Stanford University Press, 2001.

Green, Michael S. *Nietzsche and the Transcendental Tradition.* Urbana and Chicago: University of Illinois Press, 2002.

Günther, Friederike Felicitas. *Rhythmus beim frühen Nietzsche.* Berlin and New York: Walter de Gruyter, 2008.

Heidegger, Martin. *Nietzsche.* Translated by David F. Krell. 4 vols. San Francisco: Harper and Row, 1979–87.

—. *What is Called Thinking?* Translated by Fred D. Wieck and J. Glenn Gray. New York: Harper and Row, 1968.

Helmholtz, H. von. *Wissenschaftliche Abhandlungen.* 3 vols. Leipzig: Johann Ambrosius Barth, 1882.

Herbart, Johann Friedrich. *Sämtliche Werke.* Edited by Karl Kehrbach and Otto Flügel. 19 vols. Aalen: Scientia Verlag, 1964.

Higgins, K. M. *Nietzsche's Zarathustra.* Philadelphia: Temple University Press, 1987.

Horace. *The Odes and Epodes.* Translated by C. E. Bennett. Loeb Classical Library. London: Heinemann, 1960.

James, William. *Principles of Psychology.* London: Macmillan, 1918.

Kafka, F. *The Great Wall of China.* Translated by W. Muir and E. Muir. New York: Schocken Books, 1946.

Kain, Philip J. 'Nietzsche, Skepticism and Eternal Recurrence'. *Canadian Journal of Philosophy* 13 (1983): 365–87.

Kant, Immanuel. *Critique of Pure Reason.* Translated by Norman Kemp Smith. London: Macmillan, 1929.

Kirk, G. S. and J. E. Raven. *The Presocratic Philosophers: A Critical History with a Selection of Texts.* Cambridge: Cambridge University Press, 1957.

Klossowski, Pierre. *Nietzsche and the Vicious Circle.* Translated by Daniel W. Smith. London: The Athlone Press, 1997.

Lampert, Laurence. *Nietzsche's Teaching: An Interpretation of* Thus Spoke Zarathustra. New Haven and London: Yale University Press, 1986.

Lange, F. A. *Geschichte des Materialismus und Kritik seiner Bedeutung in der Gegenwart.* Iserlohn: J. Baedecker, 1866.

Lecky, W. E. H. *A History of England in the Eighteenth Century.* 3 vols. London: Longmans, Green, and Co., 1913.

Lichtenberg, G. C. *Schriften und Briefe.* Darmstadt: Wissenschaftliche Buchgesellschaft, 1972.

Liebmann, Otto. *Kant und die Epigonen. Eine kritische Abhandlung.* Stuttgart: Carl Schober, 1865.

Loeb, Paul S. 'The Conclusion of Nietzsche's *Zarathustra*'. *International Studies in Philosophy* 32 (2000): 137–52.

—. 'The Gateway-Augenblick'. In *Nietzsche's* Thus Spoke Zarathustra: *Before Sunrise,* edited by James Luchte, 91–108. London and New York: Continuum, 2008.

—. 'Identity and Eternal Recurrence'. In *A Companion to Nietzsche,* edited by Keith Ansell Pearson, 171–88. Oxford: Blackwell, 2006.

—. 'The Moment of Tragic Death in Nietzsche's Dionysian Doctrine of Eternal Recurrence: An Exegesis of Aphorism 341 in *The Gay Science*'. *International Studies in Philosophy* 30 (1998): 131–43.

—. 'The Thought-Drama of Eternal Recurrence'. *The Journal of Nietzsche Studies*, no. 34 (2007): 79–95.

—. 'Time, Power, and Superhumanity'. *Journal of Nietzsche Studies*, no. 21 (2001): 27–47.

Lucian. *Works*. 8 vols. Loeb Classical Library. London: Heinemann, 1913–67.

Macht, David I., 'A Physiological and Pharmacological Appreciation of *Hamlet* Act I, Scene 5, Lines 59–73'. *Bulletin of the History of Medicine* 23 (1949): 186–94.

Magnus, Bernd. *Nietzsche's Existential Imperative*. Bloomington and London: Indiana University Press, 1978.

McTaggart, J. M. E., 'The Unreality of Time'. *Mind* 17 (1908): 457–74.

—. *The Nature of Existence*. 2 vols. Cambridge: Cambridge University Press, 1938. Reprint, New York: Octagon Books, 1976.

Mittasch, Alwin. *Friedrich Nietzsche als Naturphilosoph*. Stuttgart: Alfred Kröner Verlag, 1952.

Moles, Alistair. *Nietzsche's Philosophy of Nature and Cosmology*. New York: Peter Lang, 1990.

Moore, Gregory. *Nietzsche, Biology and Metaphor*. Cambridge: Cambridge University Press, 2002.

Naumann, Gustav. *Zarathustra-Kommentar*. 4 vols. Leipzig: Verlag von H. Haeffel, 1899–1901.

Nehamas, Alexander. *Nietzsche: Life as Literature*. Cambridge, MA: Harvard University Press, 1985.

Neumann, Carl. *Die Principien der Elektrodynamik. Eine mathematische Untersuchung*. Tübingen: Heinrich Laupp, 1868.

Nietzsche, Friedrich. *Philosophy and Truth: Selections from Nietzsche's Notebooks of the Early 1870s*. Edited and translated by Daniel Breazeale. Atlantic Highlands, NJ: Humanities Press, 1979.

—. *The Pre-Platonic Philosophers*. Edited and translated by G. Whitlock. Urbana and Chicago: University of Illinois Press, 2001.

—. 'Time Atom Theory'. Translated by Carol Diethe (with modifications by Keith Ansell Pearson). *Journal of Nietzsche Studies*, no. 20 (2000): 1–4.

Philoponus, John. *Against Aristotle, on the Eternity of the World*. Translated by C. Wildberg. London: Duckworth, 1987.

Plato. *The Dialogues of Plato*. Translated by B. Jowett. 2 vols. New York: Random House, 1937.

Plotinus. *The Enneads*. Translated by S. McKenna. Edited by J. Dillon. Harmondsworth: Penguin Books, 1991.

Plutarch. *Moralia*. 16 vols. Loeb Classical Library. Cambridge, MA: Harvard University Press, 1927–2004.

Porter, James I. *Nietzsche and the Philology of the Future*. Stanford, CA: Stanford University Press, 2000.

—. 'Untimely Meditations: Nietzsche's *Zeitatomistik* in Context'. *Journal of Nietzsche Studies*, no. 20 (2000): 58–81.

Prior, A. N. 'Diodorean Modalities'. *Philosophical Quarterly* 5 (1955): 205–13.

—. 'Thank Goodness That's Over'. *Philosophy* 34.1 (1959): 12–17.

Quine, W. V. O. 'Events and Reification'. In *Actions and Events: Perspectives on the Philosophy of Donald Davidson*, edited by Ernest LePore and Brian P. McLaughlin, 162–71. Oxford: Basil Blackwell, 1985.

Radhakrishnan, Sarvepalli and Charles A. Moore, eds. *A Source Book in Indian Philosophy.* Princeton, NJ: Princeton University Press, 1957.

Reichenbach, Hans. *The Direction of Time.* Berkeley, CA: University of California Press, 1956.

Richardson, John. 'Nietzsche on Time and Becoming'. In *A Companion to Nietzsche,* edited by Keith Ansell Pearson, 208–29. Oxford: Basil Blackwell, 2006.

—. *Nietzsche's New Darwinism.* New York: Oxford University Press, 2004.

—. *Nietzsche's System.* New York: Oxford University Press, 1996.

Rosen, Stanley. *The Mask of Enlightenment: Nietzsche's* Zarathustra. Cambridge: Cambridge University Press, 1995.

Rosenzweig, Franz. *The Star of Redemption.* Translated by W. H. Hallo. London: Routledge and Kegan Paul, 1971.

Russell, Bertrand. *The Problems of Philosophy.* London: Williams and Norgate, 1912.

Schlechta, Karl and Anni Anders. *Friedrich Nietzsche: Von den verborgenen Anfängen seines Philosophierens.* Stuttgart and Bad Cannstatt: Friedrich Frommann Verlag, 1962.

Schlesinger, George N. *Timely Topics.* Basingstoke: St Martin's Press, 1994.

Schopenhauer, Arthur. *On the Fourfold Root of the Principle of Sufficient Reason.* Translated by E. F. J. Payne. La Salle, IL: Open Court, 1974.

—. *Parerga and Paralipomena.* Translated by E. F. J. Payne. 2 vols. Oxford: Clarendon Press, 1974.

—. *The World as Will and Representation.* Translated by E. F. J. Payne. 2 vols. New York: Peter Smith, 1969.

Schutte, Ofelia, 'Willing Backwards: Nietzsche on Time, Pain, Joy, and Memory'. In *Nietzsche and Depth Psychology,* edited by Jacob Golomb, Weaver Santaniello, and Ronald L. Lehrer, 111–26. Albany, NY: State University of New York Press, 1999.

Seung, T. K. *Nietzsche's Epic of the Soul: Thus Spoke Zarathustra.* Lanham, MD: Lexington Books, 2005.

Sextus Empiricus. *Works.* Translated by R. G. Bury. 3 vols. Loeb Classical Library. London: Heinemann, 1933–6.

Shapiro, Gary. *Nietzschean Narratives.* Bloomington and Indianapolis: Indiana University Press, 1989.

Silverman, Hugh J. 'The Inscription of the Moment: Zarathustra's Gate'. *International Studies in Philosophy* 21 (1992): 53–61.

Simmel, Georg. *Schopenhauer and Nietzsche.* Translated by Helmut Loiskandl, Deena Weinstein and Michael Weinstein. Amherst, MA: The University of Massachusetts Press, 1986.

Simpson, R. R. *Shakespeare and Medicine.* Edinburgh: E. and S. Livingstone, 1959.

Skowron, Michael. 'Zarathustra-Lehren Übermensch, Wille zur Macht, Ewige Wiederkunft'. *Nietzsche-Studien* 33 (2004): 68–89.

Small, Robin. 'Disturbing Thoughts and Eternal Perspectives: Some Uses of Symbolism in Nietzsche'. *New Nietzsche Studies* 3/3–4 (1999): 29–44.

—. 'Nietzsche and Cosmology'. In *A Companion to Nietzsche,* edited by Keith Ansell Pearson, 189–207. Oxford: Basil Blackwell, 2006.

—. *Nietzsche and Rée A Star Friendship.* Oxford: Oxford University Press, 2005.

—. *Nietzsche in Context.* Aldershot: Ashgate, 2001.

Soll, Ivan. 'Reflections on Recurrence: A Re-examination of Nietzsche's Doctrine, die Ewige Wiederkehr des Gleichen'. In *Nietzsche: A Collection of Critical Essays,* edited by Robert Solomon, 322–42. New York: Doubleday Anchor, 1973.

Spir, African. *Denken und Wirklichkeit. Versuch einer Erneuerung der kritischen Philosophie.* 2 vols. Leipzig: J. G. Findel, 1873.

Stambaugh, Joan. 'Das Gleiche in Nietzsches Gedanken der Ewigen Wiederkunft des Gleichen'. *Revue internationale de philosophie* 18, no. 67 (1964): 91–118.

—. *Nietzsche's Thought of Eternal Return.* Baltimore and London: The Johns Hopkins Press, 1972.

—. *The Problem of Time in Nietzsche.* Translated by John F. Humphrey. Lewisburg, PA: Bucknell University Press, 1987.

—. 'Thoughts on a Nachlass Fragment from Nietzsche'. *Nietzsche-Studien* 6 (1977): 195–204.

—. 'Zufall'. *Philosophy Today* 43.1 (1999): 95–9.

Stegmaier, Werner, 'Zeit der Vorstellung: Nietzsches Vorstellung der Zeit'. *Zeitschrift für philosophische Forschung* 41 (1987): 202–28.

Strong, Tracy B. *Friedrich Nietzsche and the Politics of Transfiguration.* Berkeley: University of California Press, 1975.

Teichmüller, Gustav. *Darwinismus und Philosophie.* Dorpat: C. Mattiesen, 1877.

—. *Die wirkliche und die scheinbare Welt. Neue Grundlegung der Metaphysik.* Breslau: Wilhelm Koebner, 1882.

Twain, Mark. *Jim Smileys berühmte Springfrosch und Dergleichen wunderliche Kauze mehr.* Translated by Moritz Busch. Leipzig: Grunow, 1874.

Ueberweg, Friedrich. *Grundriss der Geschichte der Philosophie.* Edited by Karl Praechter. Graz: Akademische Druck- und Verlagsanstalt, 1953.

Ulfers, Friedrich and Mark Daniel Cohen, 'The Energeticist Model of the Universe as *Perpetuum Mobile* and Nietzsche's Notion of the Eternal Recurrence of the Same'. http://homepage.mac.com/cohenmd1/.Public/Philosophy/THE%20ENERGETICIST%20MODEL.pdf. Accessed 23 February 2009.

Visser, Gerard, 'Der unendlich kleine Augenblick'. *Nietzsche-Studien* 27 (1998): 82–106.

Vogt, Johannes Gustav. *Die Kraft. Eine real and monistische Weltanschauung.* Leipzig: Haupt und Tischler, 1878.

Wagner, Richard. *Gesammelte Schriften und Dichtungen.* Edited by Wolfgang Golther. 10 vols. Leipzig: E. W. Fritzsch, 1871–83. Reprint, Berlin: Deutsches Verlagshaus Bong & Co., 1913.

Weichelt, Hans. *Zarathustra-Commentar.* 2nd edn. Leipzig: Felix Meiner, 1922.

White, Alan. *Nietzsche's Labyrinth.* New York and London: Routledge, 1990.

Whitlock, Greg. 'Examining Nietzsche's "Time Atom" Fragment from 1873'. *Nietzsche-Studien* 26 (1997): 350–60.

—. 'Investigations in Time Atomism and Eternal Recurrence'. *Journal of Nietzsche Studies,* no. 20 (2000): 34–57.

—. *Returning to Sils-Maria: A Commentary to Nietzsche's "Also Sprach Zarathustra".* New York: Peter Lang, 1990.

— 'Roger Boscovich, Benedict de Spinoza and Friedrich Nietzsche: The Untold Story'. *Nietzsche-Studien* 25 (1996): 200–20.

Williams, Linda L. and Joseph T. Palencik, 'Re-evaluating Nietzsche's Cosmology of Eternal Recurrence'. *The Southern Journal of Philosophy* 42 (2004): 393–409.

Xenophon. *Memorabilia and Oeconomicus.* Translated by E. C. Marchant. Loeb Classical Library. London: Heinemann, 1923.

Zöllner, Johann Carl Friedrich. *Über die Natur der Cometen. Beiträge zur Geschichte und Theorie der Erkenntnis.* 2nd edn. Leipzig: Wilhelm Engelmann, 1872.

Index

www.ingramcontent.com/pod-product-compliance
Ingram Content Group UK Ltd.
Pitfield, Milton Keynes, MK11 3LW, UK
UKHW020142290926
15471UKWH00004B/647

9 781441 147943